ABOUT THE AUTHORS

Dr Peter Mansfield is a family doctor with a special interest in food quality and health. Since meeting the founders of the Peckham Experiment in 1971, he has been exploring a new role for general practice in the 21st century, and in 1981 founded the Templegarth Trust to make the nature of health more widely known and to encourage experiments in cultivating it. His many articles and pamphlets include *Look Again at the Label* (Soil Association, 1985), and he is the author of *Instinct For Life*. He began a regular health column in the *Soil Association Quarterly Review* in 1980, and since 1984 has been an adviser and correspondent to *New Health* magazine, contributing to a series on food additives judged Best Campaign for 1985 by the Periodical Publishers Association.

Dr Jean Monro is Medical Director of the Allergy and Environmental Dept at Nightingale Hospital, London. She specializes in clinical ecology and the treatment of allergies, and has received a number of awards for research work. She is the author of *Some Dietary Approaches to Disease Management*.

Also published by Century Hutchinson:

Additives: Your Complete Survival Guide   *Felicity Lawrence*
An A–Z of Your Child's Health   *Dr David Delvin*

CHEMICAL CHILDREN

How to Protect Your Family from Harmful Pollutants

Dr Peter Mansfield
and Dr Jean Monro

CENTURY
LONDON   MELBOURNE   AUCKLAND   JOHANNESBURG

First published in 1987 by Century Hutchinson Ltd,
Brookmount House, 62–65 Chandos Place, Covent Garden,
London WC2N 4NW

Century Hutchinson Australia Pty Ltd,
PO Box 496, 16–22 Church Street, Hawthorn, Victoria 3122,
Australia

Century Hutchinson New Zealand Ltd,
PO Box 40-086, Glenfield, Auckland 10,
New Zealand

Century Hutchinson South Africa Pty Ltd,
PO Box 337, Bergvlei, 2012 South Africa

ISBN 0 7126 1729 9

Photoset by Rowland Phototypesetting Ltd
Bury St Edmunds, Suffolk
Printed and bound in Great Britain by
Richard Clay Ltd., Bungay, Suffolk

Contents

For our families,

and the children
this book is about.

All of them, in various ways,
went through trying times
before it was written.

Authors' Note

This book is a collaboration between two doctors who have many ideas and experiences in common, but who work quite independently of each other. PM wrote the text, and did not always have the opportunity to confer with JM on matters of opinion. In these instances he has used the first person singular. Otherwise, where expressing joint opinions, 'we' is the rule.

In Part One we set out the situation as we understand it: how chemical pollution began, its extent, the effects it has on its victims, and why doctors have been so slow to recognize these.

Part Two looks at a wide range of particular cases which substantiate our claims. They are grouped so as to illustrate the mechanisms involved, and told as far as possible verbatim from correspondence and other records. All of the histories quoted are genuine, and mainly come from JM's practice records. All the names have been changed, and the cases are quoted to illustrate the kind of problems that arise commonly. Reference to any particular line of treatment, whether successful or not, has been deliberately minimized. It is not our intention to advocate, nor to criticize, any particular treatment in principle. No two cases are ever alike; each requires its own individual management, worked out in consultation between the patient and his or her chosen advisers.

Part Three sets out everything sufferers and their parents can easily do for themselves. That includes drawing on the help of their existing advisers to the maximum mutual benefit, where necessary helping the advisers to change their viewpoint and attitudes to these conditions. It is quite practical to turn directly to Part Three and to ignore the rest of the book. But in a highly controversial and emotive field we feel bound to account for what we suggest, knowing that in any case many people think deeply about the meaning of their illnesses, and appreciate full and sympathetic explanation of them, provided in Parts One and Two.

In writing the book at all, our fundamental purpose lies deeper. Both of us are convinced from our combined clinical experience that real, massive and serious problems exist. We wish to share our experience with you. We are unlikely to convince the scientific establishment of the validity of what we say because we have neither the resources nor the position from which to do so. But we believe we understand something of what is going on and have some promising suggestions about setting matters right.

If what we have to say accords with your own experience as parents and citizens, then it will help you understand how that experience came about. You may go on to discover solutions to problems you now have, which will help to confirm your understanding. You will then have the confidence to pass on that insight to your friends, your professional advisers, your political representatives and the shop managements you patronize.

We have already come a way along the road towards acceptance. We are past the stage of being ignored, but have not yet cleared the hurdle of ridicule. How rapidly we pass on to general acceptance, appropriate research and political action depends on what the general public make of the situation.

You – readers, consumers and voters – must decide.

PM & JM

PART ONE

THE CHEMICALS

CHAPTER ONE
Just What is Happening?

Jack

'I had been on a wholefood diet for about a month and felt much better in myself, and my eczema was slightly better.

'I was working for my dad. I felt a bit hungry and as I did not have any of the food of my special diet with me, I went to the nearby supermarket to see what they had. I looked around and could not find anything wholemeal, so being hungry I picked up a pack of six white rolls thinking they would not be too bad for me. I took them back, filled them with ham, and ate them.

'Within fifteen minutes I had started to scratch myself badly. My dad looked at the packet and said, "Do you realize how many E-numbers there are in these rolls, Jack?" There were eight. For about an hour I was in great discomfort and couldn't work. I first didn't know what to do to help myself. I stood up, sat down, bashed my head against a wall, tried to get some fresh air. Nothing helped – I have never felt so bad. I won't do that again in a hurry.'

These were the contents marked on the packet of Jack's rolls: flour, water, yeast, vegetable fat, whey powder, salt, emulsifiers E471 (glyceryl monostearate), E472(e) (mono- and diacetyltartaric acid esters of mono- and diglycerides of fatty acids), dextrose, soya flour, preservative E282 (calcium propionate), flour improvers E300 (vitamin C), 920 (l-cysteine hydrochloride), 924 (potassium bromate), 926 (chlorine dioxide), 927 (azoformamide). And the flour contained another long string of additives before it reached the baker! (See chapter six.)

This story is a very clear example of a medical problem which is only about fifty years old. It is now affecting a large minority of young people throughout the developed world – estimates vary between ten and thirty per cent. Children, teenagers and young adults in their twenties are the first generation to have grown up totally cocooned in chemicals. Each new year's babies encounter a stiffer challenge to their physical constitution than was faced by older children in their day, a year or two before. The numbers of children adversely affected by chemicals now amount to an epidemic which ought to be obvious, recognized, taken seriously and dealt with.

On the contrary: its existence is denied or ignored by the majority of the medical profession, and most government officials, in every affected country. They consider it an illusion, mass neurosis, an alarmist exaggeration – anything rather than accept it at face value. Why?

To begin with, doctors may have been in practice twenty years already without yet recognizing what we claim to see. Since their ideas were formed on the basis of experience available twenty or even ten years ago, they may now be well out of date. Naturally doctors do not change their minds lightly, and their reluctance precedes them in their reputation. If people find such doctors unhelpful, they do not consult them again. These doctors are amply preoccupied with work they do well, and give the problem little further thought.

We see things differently, challenged repeatedly by the problems people bring to us. We believe this epidemic can be accounted for in perfectly scientific terms. And the resistance put up by governments and established medical authorities since the 1940s is all part of the story, which begins with how medical theories of allergy have developed.

Ideas about Allergy

Allergy has always been understood as an excessive response to a reasonable provocation. Grass pollen gets up everybody's noses; only a few respond with hay fever, a third of one per cent. A few people show similar exceptional responses to other pollens, or to consumption of particular foods, such as shellfish or strawberries. Insect stings and animal fur are other well-recognized provocators, or allergens. Allergy is regarded as one of the causes of asthma, which like hay fever is a stock reaction to inhalation of one or several allergens, though it can also be provoked in other ways.

A few children – about five in every thousand – can be found with one or more of several conditions, all thought to be allergic in origin: asthma, a particular kind of eczema and perhaps hay fever. These 'atopic' children often have parents or other near relatives who have suffered in the same way. They represent the accepted picture of juvenile allergy in its extreme form. So allergy, in the old textbooks, was understood as a family of conditions with excessive response as their common theme. Taken all together, they affected no more than about four per cent of the population of Britain.

Then laboratory immunology was invented. Doctors began to study the biochemistry of allergy, and discovered a number of things happening in the bodies of allergic people which distinguish them from everybody else. At certain times they have much larger numbers of one kind of white cell circulating in their blood, particularly when experiencing severe allergic reactions. And they build up vast amounts of protein in their bloodstream between reactions too. These are the antibodies we all have, but allergic people produce many more than they need. These findings can be understood as gross exaggerations of the two basic kinds of immune response available to normal people – the 'cellular' and the 'humoral' mechanisms, as they were first described. The tale of discoveries after that is complicated and confusing, and to this day provides endless opportunities for speculation.

Our main concern is how immunologists redefined allergy. At a time when biochemical precision was all the rage, the definition by patterns of symptoms went out and a new list of criteria came in. Only a full range of positive blood tests and clear demonstration of specific allergens, which always give the same effect and which cannot be faked, make a confident modern diagnosis of allergy – a diagnosis which continues to be uncommon.

In fact, there is only a tiny minority of children with symptoms of the allergic type whose problem fits this diagnosis. Their bodies overreact hugely when exposed to a small range of natural substances familiar to all of us, including some things most people consume as food with perfect safety. But there are a great many more children with similar symptoms, who fail to show positive results to these tests. Accurate statistics would be very difficult to collect, but there seem to us to be at least four or five times as many children in this predicament, and the numbers are rising steeply. Where do these children stand?

Coping with Stress

Theron Randolph, a Chicago physician and clinical scientist, spent several decades working on this question. In the experimental work of Professor Hans Selye of Montreal he discovered an explanation for some puzzling symptoms he had observed in his patients. From this he was able to build a respectable theory for the basis of their disease. Many physicians in America have put this successfully into practice, and some, like Ben Feingold, have extended the theory in useful ways.

Richard Mackarness was one of the earliest English physicians to meet Dr Randolph. He applied the same theory to psychiatric patients under his care in Basingstoke District Hospital. He became famous ten years ago with the publication of his results in popular form, having failed consistently to impress the British medical press. The success of his books probably alienated the medical establishment further and helped to fix the rift in medical thinking in this country which has persisted ever since.

Dr Randolph's theory was based on a series of bodily responses to stress, known as the General Adaptation Syndrome. Selye observed it originally in rats, but it can be seen equally well in children. The syndrome begins with a child's repeated occasional exposure to any irritant substance. At a certain minimum exposure level, perhaps after several repetitions, a perfectly understandable rejection response occurs, appropriate to the circumstances. This is the 'alarm' response, and will be reproduced whenever that child is exposed to that substance boldly enough, provided this only occurs occasionally.

If the exposure happens frequently or goes on all the time, the pattern of response changes. The child 'resists' the effects of the substance, and appears to be adapted to it. However, this costs the body a considerable internal effort, making much larger quantities than usual of hormones such as cortisone and adrenalin. Their presence in the blood can give the child a boost of energy and well-being which he begins to miss after an interval, when the hormones are spent. The child may therefore develop a special liking for the substance, amounting to an addiction which must be satisfied at regular intervals. If it is not, then the adaptation breaks down and the next exposure will cause an 'alarm' response once more, whose violence will depend on the interval since the last exposure. It may after months of abstinence take repeated doses of exposure to re-establish the 'alarm' response.

If in the 'resistance' phase exposure is maintained, symptoms may not appear for weeks or months. But they show eventually, once the body's adaptive ability breaks down. This is the phase of 'exhaustion'. The boost no longer occurs because insufficient cortisone is available, and the child feels ill or irritable permanently. Withdrawal may at first make him iller, but it is the only way to rest his system, end the addiction and restore normality. Unfortunately, the whole cycle can be rekindled at any time by renewed exposure. All this can happen without any of the formal criteria for allergy ever being fulfilled. This is not a disease peculiar to the particular child, but a system of normal responses we all have to stressful circumstances as our ability to cope with them diminishes.

A wide range of symptoms in particular people, including the well-known allergic responses but extending into behaviour, thinking and personality, can be explained by this theory, and traced to causes in one or more common foods or pollutants of water or air. At first Mackarness's cases were often adults, whose 'resistance' phases were generally longer and who could cooperate better in his enquiries; but once the phenomenon was recognized it became clear that children were commonly and often seriously affected.

Rachel Carson had meanwhile called attention to widespread creeping chemical pollution that had begun somewhere in the 1940s and had penetrated many kinds of plant, insect and animal. Her book, *Silent Spring*, chilled its readers: if this process had already reduced the population of birds, whatever might it be doing to people?

McCarrison (a British Army doctor) in the 1920s, and Pottenger (an American dental surgeon) ten years later had already proved, from observations in humans and in animal experiments, that polluted or degraded food is the chief cause of many degenerative diseases in human beings. The answer that

Randolph and Mackarness gave is more immediate: chemicals, when consumed in food or drink or inhaled, can quickly produce allergic-type responses in children and adults alike. These responses are often very dramatic and impressive to witness, but lack that important feature – positive results in the accepted biochemical tests for allergy.

A New Idea – Intolerance

Allergy, in its original and true sense, remains as uncommon as ever it was. Mercifully, few people are programmed by nature to react massively and unreasonably to life's ordinary challenges. But a massive response may not always be unreasonable. If food, water and air now contain quite unnatural chemical substances, even in very small quantities, might not they prove very irritant to quite non-allergic people?

The incident with which we began was described by a teenager with eczema who had begun to improve after some radical changes in his diet had reduced appreciably his exposure to such chemicals. This reduction had begun to rescue him from the 'exhaustion' phase of adaptation, represented by general ill-health and eczema.

Accidental re-exposure to these chemicals set him off severely in 'alarm' phase symptoms, more acute than he can ever remember having before. These are indistinguishable from the reactions of classical allergy. But he is unlikely to show clear and reproducible biochemical reactions to any of the chemicals he met in his buns.

Paul's experience provides another simple example of a different kind.

Paul

10 December

'Paul has severe catarrh which is provoked by milk and chlorine [discovered by techniques discussed in Part Three]. He is also subject to recurrent mouth ulcers, particularly if he eats beef.

'He has very large tonsils, and catarrh dripping down his throat from the back of his nose. In the past he was diagnosed as having nasal polypi, but was not keen to have a polypectomy.'

His parents were, as a first step, advised how to avoid bleaches and household cleaners, how to treat bath water and filter water for drinking. That dealt with the chlorine, an obvious irritant to which intolerance is not unreasonable, leaving the question of allergy to milk for later attention.

Ten months later . . .

1 October
'Paul has done very well on the very simple regime of treatment for chlorine hypersensitivity which has been completely successful in controlling his symptoms.'

The outcome of this story illustrates another feature of such situations. If Paul avoids chlorine, he can cope with beef and milk in moderation without ill effects. Yet he originally reacted as if he was allergic to them. Had his intolerance of chlorine not come to light, he might easily have been type-cast on clinical grounds as allergic to the cow; yet biochemically, there was no evidence of this. He can switch his tolerance of beef on and off, depending on whether or not he is being irritated by chlorine.

According to Hans Selye, we all tolerate a whole range of low-grade irritations of this kind up to a certain point. But as they accumulate during our lives, or as the total concentration of irritant chemicals present in our food and environment increases, a point is reached where our tolerance breaks down. When that happens, the symptoms produced are, like Paul's and Jack's, identical with those seen in strict allergy.

Intolerance is a better word than allergy to describe this kind of breakdown. It is the reasonable response of normal people to intolerably abnormal circumstances. It represents healthy bodies in self-defensive action, and is nothing at all to do with allergy. So we should not expect the tests for allergy to be positive. It is not people who are 'abnormal', but the circumstances! This clears up a confusing situation, and lets allergy retain its classical, restricted meaning. We shall refer to reasonable reactions against obvious irritants as intolerance, to distinguish them from the unreasonable reactions to familiar natural substances that we call allergy.

Is intolerance new, or just newly recognized? Almost certainly we have underestimated it up to now; when doctors start to look for it, they recognize many cases that puzzled them before. But it seems also to be increasing rapidly as a problem in our time. More than six million chemicals are known today, and the number in common use has rocketed since the middle of this century. Something like a thousand new kinds are being invented and introduced every year. It would be surprising if these did not irritate many other people, who are provoked into producing the same reactions as the minority, who have always been allergic to more familiar things.

We believe this is exactly what is happening. It accords with our experience as doctors, which is the subject of Part Two of this book. And it corresponds with the wide range of trends we believe cause that experience, which are detailed in the next five chapters.

Overload and Incredulity

We have yet to explain why the medical profession is so resistant to this idea if it is so obvious to us. Reliance on behavioural observations in people has certainly been a major stumbling block. Though perfectly scientific, such methods are currently unfashionable. They are further discredited by the unbelievable results they appear to give.

Allergies come in ones and twos, and are relatively permanent; intolerances come by the dozen, and are very variable. Once your tolerance of stress begins to crack, it soon breaks down completely. In Paul's case, active chlorine intolerance unmasked intolerance of milk.

This tendency to crumble on all fronts once the defences have been breached is termed 'overload'. It is a prominent feature of intolerance, and would be unthinkable and unmanageable as a feature of classical allergy. Confusion of the two by enthusiasts, and their attempts to deal with large numbers of intolerances by classical desensitization methods, makes unconvinced doctors quite reasonably incredulous.

Nevertheless, the problem does occur and has to be faced. Peter's sensitivity started when the house was proofed against woodworm and insulated with urea-formaldehyde foam. When his room was renovated new chipboard flooring was laid, which gassed off formaldehyde. These several chemical irritants broke down his general tolerance: he became intolerant of a whole range of foods, and in his adaptation to sugar, craved it like an addict (see next page). Let the record speak for itself.

Peter

28 August

'Peter's abnormal behaviour is very much tied up with his craving for sugars. He had a normal childhood and was noted to be bright and happy, but he used to have asthma and rhinitis. At the age of eight years it was noted that he was not achieving very good results at school and therefore he was moved to another school; but the master reported that he was "totally destructive", and said that he could not cope with him. It was also noted that his craving for sugars was so great that he would break into other students' lunch boxes. Eventually he was suspended from school.'

Under test, most of his reactions were to fruits and vegetables. When challenged with sugar and chocolate he did not react, because he had not had these for the previous three months and had recovered his tolerance of them. However, when tested with fructose he suddenly became extremely hyperactive: it was clear that this explained his reactions to fruit and vegetables. He was tested to a wide range of other foods, chemicals and inhalants, and as a result was offered immunotherapy.

Immunotherapy was invented in the 1970s by an American allergist called Miller, who found that allergic symptoms provoked by challenge with food substances could then be neutralized by diluted solutions of the same substances, either as drops in the mouth or as injections. This has formed the basis of a laborious but useful testing and treatment method offered by clinical ecologists, in conjunction with nutritional analyses of blood and hair, systematic rotation of menus, nutritional supplements and the removal of irritant chemicals from the patient's environment. Practical details of these methods are described in Part Three.

In Peter's case, advice was given on the chemical clearing of his home and on a rotating diversified diet. His mineral status was poor for magnesium, for which he was given a supplement.

19 October

'Peter is getting much better. He is more alert mentally as well as physically and according to him, he is now starting to enjoy life. His mother has his diet well controlled. He has also moved school and his head teacher is pleased to note that his behaviour is normal. He is now taking his treatment three times a week.

'His blocked nose returned two weeks after he left hospital and is mainly due to the chemicals in his room. His parents are planning to move him to another room which they are in the process of converting.'

4 January
'Peter is extremely well. He has had his treatment, and been on a four-day rotation diet. There have been no behavioural disturbances at all and no asthma. He has also been growing rapidly, but he continues to be a little tired.'

At first glance, overload symptoms like Peter's automatically suggest exhibitionism, over-indulgence, neurosis or plain bad parenting and are hard to credit at face value. The fact that a child like Peter can get his act together, rebuild his tolerance and start to live again is an impressive personal testimony. Of the doctors and public figures who now understand the problem, almost all became convinced of the efficacy of this approach when they saw it being worked out in their own relatives.

Otherwise, it remains very unattractive to doctors accustomed to ideas which can be tested by concrete evidence. If we claim that their usual tests need not apply, that provocation by a particular irritant need not necessarily always result in the same symptoms of intolerance, and that irritants may substitute for each other, we are asking them to accept a package of vague uncertainties. Even straightforward chemical analysis of the mineral content of samples of hair remains controversial, because the implications of the results completely overthrow conventional views of mineral nutrition. Some of the new tests that are proposed – like Miller's provocation/neutralization test, results from which will be repeatedly referred to in Part Two – depart so far from conventional chemistry and physics that they leave established doctors openly incredulous.

Michael is a boy with formidable environmental problems. At one stage in his progress he proved to be intolerant of thirty-seven foods, tap water, house dust and house-dust mite and cats. His parents consulted all sorts of doctors, with varying results. Here are two contrasting points of view, expressed by different specialists, on the subject of whether or not he was short of nutrient minerals, particularly zinc. First, some observations by a well-established conventional physician.

Michael

18 January

'Michael was reviewed here on 20 December when I was pleased to note the improvement in his behaviour after immunotherapy from a clinical ecologist. He had a hair sample analysed for minerals and I gather that this shows that many of these (including zinc) were low. The question of malabsorption has also been raised as a possible cause of his symptoms.

'I therefore took a blood sample and measured some of these. The following minerals were within normal limits: sodium, potassium, calcium, magnesium, copper, zinc. Blood lead level was below the normal limit allowed in children.

'I was therefore unable by this method to document any significant mineral deficiency, or any evidence of malabsorption as related to blood count. I have therefore not suggested any further intervention in this regard at the moment, but have written to Michael's mother to reassure her that Michael's blood mineral levels are within normal limits.'

The first doctor's opinion was reached competently with reference to techniques he felt he could trust, but it did not help Michael. Had he abided by this advice, his health would probably have deteriorated further.

See now the comments of a medical specialist in nutrition, still a rare breed, later in the year.

6 June

'I requested a repeat hair mineral analysis to assess the degree of improvement following the supplementation programme. Of most interest was the fact that his toxic metal profile was . . . really quite excellent.'

'*Comment on previous findings*. A low hair zinc is indicative of inadequate zinc nutriture. Blood serum is about the last tissue compartment to have a reduction in zinc levels. The fact that his hair zinc was low really indicates that his nutriture was inadequate over the three months prior to the sampling of hair, even though the serum value was normal.

'I shall be seeing him again in a couple of months' time and will drop you a note following that visit. In the meantime, he continues to take his nutritional supplements, including evening primrose oil, and if his overall symptomatology is the same in a couple of months' time, I would recommend adding (to this).'

This doctor's comments clearly represent a different and more positive point of view. It explains why the first blood zinc result could not be trusted, gave Michael's parents' clear guidance about what to do, corroborated the previous hair test, and demonstrated the later improvements in return for all their efforts meanwhile. When managed along these lines, Michael continued to improve, though slowly.

Just as hard for conventional doctors to believe are the effects that responses of intolerance are claimed to have on human behaviour. Randolph, Feingold and Mackarness for years drew attention to the phenomenon of hyperactivity in childhood, and demonstrated time and time again to eye-witnesses in their clinics the validity of this phenomenon. Hundreds of parents can testify to it for themselves, and to its dramatic resolution in their child when the cause is discovered and eliminated.

To believe in hyperactivity, you have to see it. There is no laboratory test yet devised which would impress a sceptic at second hand; he has to be prepared to come and sit in a clinic and be shown. That involves a major step away from establishment ranks, and requires personal courage. It is uncomfortable work, with all one's assumptions and their scientific basis under question.

Michael's experience again illustrates this clearly. Here are two comments on his aggressive behaviour. The first is from a highly respected clinical professor at a London teaching hospital. The letter was written by his assistant, also a competent paediatrician.

6 December
'The Professor and I saw Michael together today, in view of the most unusual history of violent and disturbed behaviour in certain environments and in consideration that this may be an allergic phenomenon. He is certainly relatively immunodeficient, and it is likely that the diarrhoea and alimentary symptoms he experienced earlier were of food allergy origin. I am attempting to broaden the diet. I suggested the use of Disodium Cromoglycate [a drug which blocks an allergic reaction], and his behaviour has altered remarkably since this drug was given. His mother has even taken him to places that apparently exacerbated his symptoms in the past, without effect. I

know of no description of such an effect for this drug and, clearly, the interpretation is highly subjective, but I think the implications are sufficiently profound that we should try an intermittent use of the drug and placebo over a course of a few weeks to see whether there is an effect.

'The Professor assessed Michael as behaving as a normal boy when he saw him today, but the history certainly supports a story of behaviour disorder. There are, of course, a number of background factors that could be alternative explanations for this, but the use of placebo will provide us with a clear answer in this case.'

The paediatrician is clearly having difficulty digesting the situation at face value, and is trying to fit it into models he is comfortable with. As he cannot, his next priority is scientific scrutiny of the apparent effect of the drug. This will take his mind off the uncomfortable and fundamental questions for which his profession provides no answers.

In contrast, here is the simple and detailed eye-witness account given by Michael's nursery school head teacher, just over a year later.

29 January

'Since Michael has started at nursery school his behaviour has always been unpredictable, but towards the end of last term he became noticeably more "frantic" in both movement and behaviour. He seemed not to be able to judge distances or obstructions at all. For example, if a group of children were sitting in a cluster and Michael wanted to pass them, he seemed incapable of by-passing them – he simply crashed straight into them, truly without being able to help it. This might happen half a dozen times a morning. He would also move in a very jerky fashion, without any coordination or pre-judgement.

'He moves faster than any other child I have ever dealt with but in a very frenetic way. When he's like this he's not a very happy child, it's almost as if he's using all his energy in a short period of time, his physical energy, which then upsets him emotionally. He simply cannot sit down and concentrate on anything, and he unintentionally disturbs the other children next to him with his very wild hand and arm movements.

'Both his mother and I felt that perhaps our radiators were seriously affecting his behaviour. They are ordinary electric radiators which are covered with a wooden frame and mesh wiring. Unfortunately, it is very easy for the dust to be trapped inside. The dressing-up clothes we felt might also be to blame since they're worn continuously by twenty to thirty children, and Michael loves dressing up. These clothes are now washed after five days' use and I feel Michael's behaviour when dressing up is on a par with the other children.

'The radiators were taken apart and thoroughly cleaned, then washed with Dettol, but I find it very difficult to tell whether it's had any lasting effect. One day Michael can behave well, the next he may be frantic, which must mean he's being continually irritated by something.

'A fairly typical bad day for Michael would begin at 9.30 am when he arrives. He'll rush for the dressing-up clothes and literally race from one room to the next. By 10.15 am we all sit down to sing and Michael will plough through all the children to find a place, then he'll move at least half a dozen times, under the tables, on top of another child, anywhere in fact where there is no room for him to be!

'Then when the others do a tracing or cutting-out, Michael will try because he's genuinely interested in the activities, but he just cannot concentrate, he has to be continually moving. He loves going in the garden because unless he's a danger to others, his movements there are unrestricted. When Michael behaves in this frenzied and "hyperactive" way I feel he really is at odds with himself.

'I apologize if this letter is rather disjointed but it's extremely difficult to both write and think with all the noise of the whole nursery.'

Free from pre-conceived notions, this man is able to give acute sympathetic and highly informative observations. He has no answers either, but does not feel under pressure to provide them. Doctors are uncomfortable not to be able to.

The Political Dimension

The issue of allergy has tended to divide exactly along the lines you might expect, with younger doctors more inclined to recognize and explore the need for change, and established doctors sheltered from it by their preferences and reputations. And since government generally draws its expert advisers from the established section of the profession, politicians tend not to see the problem either. This enables them to deny the need for action, which evades the issue of how difficult that action will be. They are encouraged to maintain the status quo by powerful pressures, generated by massive industrial vested interests, the subject of the next five chapters. While experts continue to support these interests indirectly, any need to face and overcome those pressures is postponed.

The problem does not, however, go away. We need to break the deadlock, which our elected representatives, the food industry and the medical establishment seem unable to accomplish. How is it to be done, if not by political process based on laboratory science?

Domestic science may provide the answer. It is in the home that the problems are most vividly manifest. It is domestic expenditure which buys the foods and chemicals we have doubts about. And it is from home that voters set off to the polls. If there is a problem, and the great mass of the British public can agree that there is, then pressure sufficient to overwhelm any vested interest can be mobilized, and will be rapidly successful. Politicians in general may not always lead well, but they are quick to follow a genuinely popular trend.

CHAPTER TWO

Chemists Down on the Farm

Vast tracts of Britain were once covered by virgin forest; man's influence has today made such forest rare. Our increasing numbers have exerted steady pressure on the earth's resources of food, so that hunting and gathering what was naturally available ceased long ago to provide for us sufficiently. So down came the trees, clump by clump. We began to cultivate, making more and more intensive use of the land, and steadily increasing its yield of crops we could consume.

Agriculture has never been better than a compromise between the needs of people and of the natural wilderness they settled in. But it has mostly been on nature's terms. All our wisdom and collective experience, throughout prehistory and well up into our recorded past, has held us in awe of natural law.

Long before religious doctrines could be defined, natural reverence became the rule. While men were weak and obviously at the mercy of climate and calamity, they had no choice but to bow to superior forces. It cannot have occurred to them to rebel. With what could they conceivably have rebelled?

So they cooperated. What the soil needed, the soil received. Due amounts of rest, variety of cropping and feeding with the composted excreta of animals and household therefore became the rule. The soil was nursed and nurtured, subject of silent worship. And so it remained until the Industrial Revolution made us bold and over-confident.

Chemical Change

The early nineteenth century was an exciting time in Europe, with technological developments and scientific discoveries tumbling over each other daily. It is easy to imagine why people lost their sense of perspective, even their heads. Nevertheless, the naivety of the events which gave birth to today's radical departure from agricultural tradition is still a breathtaking embarrassment.

Much of the foundation of modern chemistry was laid when lecture-halls resonated to the pronouncements of men whose names still carry weight with school examiners, generations later – Boyle, Lavoisier and Leibig. To most O level students of chemistry, the name Leibig recalls only his invention of the glass apparatus with which they struggle to condense vapours from their experiments into acceptable droplets of liquid. This hardly does justice to his accomplishments, but perhaps it is better so.

Baron Justus von Leibig was a much respected and highly prolific figure in the scientific world of his day. We have already seen how creditable new ideas from unknown minds are slow to be accepted. It can also work the other way. Dubious theories on famous lips command more respect than they deserve.

So it turned out with Leibig's most celebrated experiment. With his interest in the chemical composition of soil, he set about investigating it just as he must have done many other materials: he burnt it. Perhaps he had not yet invented his condenser. Or perhaps, overwhelmed with the difficulty of analysing the smoke, he did not bother to use it: after all, organic chemistry had yet to be invented. For whatever reason it may have been, the smoke which arose from his burning samples of soil made no further contribution to his experiment.

Exercising the simple analytical techniques available to him, Leibig identified three principal elements in the ash of the samples of soil he had burnt. These were nitrogen, phosphorus and potassium. No matter where he collected his samples, of whatever kind of soil, the same three elements were found consistently.

On the basis of these results, after due reflection, Leibig published in 1840 an essay which was to alter agriculture beyond recognition. Under the title 'Chemistry in its application to Agriculture and Physiology', it suggested that the nourishment of plants was purely a matter of providing the simple chemical substances which can be identified in their ash, and which could be found in abundance in the ash of the soil they grow in.

To be fair, this was consistent with the confused theory of the nature of life generally accepted at the time. It was to be another eighteen years before Antoine Béchamp, the genius of nineteenth-century French biology, exploded it, with Pasteur in his wake. Nevertheless, it was a very crude piece of reasoning.

Perhaps if things had rested there, peasant lore would quickly have prevailed to bury this nonsense decently. It was the practical application of Leibig's experiment which prevented it. For, when soil was dressed with mineral fertilizer rich in nitrogen, phosphorus and potassium, its yield of crops considerably increased. Not only did this appear to justify Leibig's conclusions in the only legitimate way, by experimental confirmation of predictions based upon them; it also promised enormous commercial advantages. Profit can seldom before have been so openly justified by respectable scientific results. It was only the conscientious misgivings of a few, the conservatism of traditional farmers and, above all, the practical limitations of the time which conserved the old agricultural ways for another hundred years.

Surrender

Two major developments during this century finally overthrew tradition, and established chemistry in its place. One was the growth of organic chemistry, coupled with increasing exploitation of the chemical wealth to be found in wood, coal and crude oil. The other was mobilization of the chemical and heavy engineering industries for production of war materials.

These developments prepared the ground for the internal combustion engine, which rapidly went into quantity production under the stimulus of the First World War. Shipping developed swiftly into its coal age, making world trade in bulk commodities much more economic. Mineral resources far afield were therefore made available to the growing industrial plants of Europe and America. The special value of nitrates in explosive manufacture gave these ores a prominent place in cargo manifests.

The Armistice in 1918 set back these developments, but left resources and machinery available for other kinds of exploitation. Many of the minerals being imported in large quantities were a suitable basis for fertilizer manufacture, and the factories which in wartime produced lorries and tanks soon turned their attention to tractors. Swords were indeed becoming plough-shares.

So began the modernization of Western farming. It was not a question of whether we needed it, but a simple exercise in marketing. We were taught to require the material resources that happened to be available. Industries stimulated by the necessities of war could not be allowed to die, for fear of economic consequences too awful to contemplate. They happened in any case, after some delay: the unemployment and collapse of land values seen in the 1930s owed something to the indigestibly rapid mechanization of the previous decade.

The Second World War rescued the economic situation, with ramifications far more diverse. By 1945 we had sophisticated radio communications, electronics, plastics, a wide range of other synthetic petrochemicals, and the beginnings of biochemistry. The biochemists in turn produced the first manufactured drugs, amongst which penicillin had already done impressive service in combating lethal infections of war wounds. Chemical engineers, however, made the largest contribution. To sustain the massive bombing campaigns which devastated the cities of Europe, enormous supplies of nitrates had been developed and stockpiled.

This time farmers came under great pressure to adopt the use of chemical fertilizers, and eventually all did, save a tiny minority. There followed a confusing array of other chemicals, which continue to baffle and disturb their users today. But the companies which sell them also employ many of the bright young graduates from agricultural colleges and the science departments of universities to research and develop their product range, and to give their customers the necessary technical advice. Many have now become wholly or partly dependent not only on agrichemical products, but on policy and management assistance from their manufacturers.

The result of this trend has been the growth of agribusiness, a completely technical and financial approach to farming – a denial of everything the traditional farmer has held to be precious and fundamental and a flagrant violation of natural law.

From Soil to Spoil

When Leibig burned his soil samples, the multitude of living creatures which had inhabited it went up in smoke. Most were microscopic, but that was really no excuse because many scientists had already taken advantage of van Leeuwenhoek's invention of the microscope and had reported masses of strange visible globular forms. But no one understood them, and 'microzymas' were not explained until Béchamp gave an account of them in 1864.

Now we know that these, and microorganisms composed of them, constitute the characteristic of healthy soil, and distinguish it from the mineral powder Leibig supposed it to be. There are staggering numbers and varieties of microorganisms, yet they weigh in as a minority component. Nevertheless, it is the microcosm they create for themselves which gives soil its distinctive structure, stability, physical properties and the nutritive quality from which all other life ultimately stems.

Good soil forms crumbs, rather than fine powder. These crumbs are held together by the fine web of organic matter fabricated by the living microcosm, which in turn has two contrasting properties. On the one hand, crumbs retain moisture as effectively as a sponge. On the other hand, they do not easily become waterlogged in a heavy fall of rain because the drainage they provide is also excellent. Similarly, air is able to penetrate the soil to a considerable depth without drying it out, partly because of the porosity between the crumbs and partly because of the network of tiny passages created by the insects and earthworms which invariably inhabit rich soil. These will have had the sense to run, long before Leibig could strike a match!

The microorganic tissue of soil has two further vital properties. Fungal elements within it are able to form a kind of web called a mycorrhiza, which surrounds and penetrates into the structure of plant roots, enabling them to draw more bountifully on the nourishment available. This and other soil organisms can even ration the nutrients, carefully locking them up in insoluble organic bonds out of season so that they cannot be washed away, and releasing them into solution for uptake by the mycorrhiza just when they can be of most use.

This automatically allows for the special peculiarities of each particular season – abnormal temperature, humidity and so on.

This remarkably sophisticated civilization thrives on the variety of the herbage rooted in it, which traditional farmers provide by rotation of crops. It conserves and consumes the composted remains of the dead organisms within it, reinforced by the fallen vegetation and excreta on its surface. Provident farmers ensure regular dressings of well-rotted farmyard manure or sewage sludge to maintain this nourishment in continuously cropped soils. Tended with surface cultivations by shallow digging tools, such soil can increase its sophistication, depth, fertility and stability almost indefinitely, demonstrating the most amazing resistance to erosion and leaching over a wide range of climatic conditions.

When we start to apply chemical fertilizer, all this changes. At first it is over-stimulated, and crop yields increase; but the health of the soil suffers. The microorganic tissue is damaged, and many of its members die off. The mycorrhiza becomes less luxuriant and shows evidence of decay. Worms and insects are less numerous, migrating from fertilized to unfertilized land where they can. The organic bonds which conserve nutrients are disturbed, and the crumb structure of soil particles degenerates. In effect, the increased crops of one year are paid for by drawing on the bank of fertility stored up for succeeding decades. As the chemical fertilizer effect drops off, more of it is needed to produce results anywhere near those enjoyed in the first season it was used. And a complication begins to show: the crop plants become weedy, less well favoured, and much more susceptible to disease.

This process is entirely natural, and familiar to the peasant farmers of ages past. As the soil weakens and loses its health, so do the plants dependent on it. Nature deals with weak creatures by making them food for the strong, so that the vigour of surviving plants and animals is continuously maintained and developed. These stronger creatures, mainly insects, fungi, bacteria and opportunistic plant weeds, would overwhelm and destroy the weak crop if left to themselves. In this way the natural vigour of the life in the neighbourhood would be restored. The peasant farmer would realize the lesson of these developments, and abandon his experiments with bags of nitrogen, phosphorus and potassium.

But there was an alternative – to depart from nature even further. In the course of exploring the properties of the chemicals being discovered in and invented from wood and fossil fuels, certain highly toxic substances were discovered. Some of these went into use in the trenches as nerve gas, and were subsequently banned by international treaty. But what we had been prepared to do to each other we soon thought of trying on other forms of life.

If your weak crops are being overwhelmed by stronger species from the uncultivated parts of nature, you could save them by weakening the strong to the level of your crops. It was tried, inevitably, and it seemed at first to work. Nature was set reeling, in retreat. Chemists congratulated each other, feeling like the saviours of mankind. Farmers were uneasy, but unsure what else to do. Ordinary people had no idea that it was happening at all.

The disordering of nature had now really begun. From Rachel Carson we learned about the accumulated devastation caused by DDT. Its use was in due course officially forsworn, but it was still obtainable on the black market in 1985 and was detected in several samples of fresh produce tested that year by government chemists.

DDT is only one of the multitude of herbicides, insecticides and fungicides which have poured into use in the last few decades. Numbers of all forms of life have declined drastically in consequence, making the overall ecological balance very unstable. So plagues of infestation can no longer be resisted – except by serial applications of yet more of the chemicals which are causing them.

Meanwhile soil now has to be cultivated deeply and drained artificially to maintain its aeration and combat waterlogging. Crumbs are a thing of the past. Fertilizers run into the drains with each flush of rain, taking with them minerals released mistakenly from bond into solution by a grossly depleted soil microculture in total confusion and disarray. Residues of pesticides intended for plants above the surface wash through below it, dooming what little life remained there and in the watercourses fed by the drains. We have at last, after a century of effort, succeeded in reducing some soils to what Leibig originally thought they were – mineral mud and dust. That dust is breaking up and blowing about the world, unable any longer to resist even the gentlest climatic pressure.

The Chemical Calendar

Consider the kind of spraying programme now required to keep a mixed farm going in England. Many variations are possible, and the following chart does not illustrate the most intensive. We have left out the dressings of fertilizer and chalk which will also be used during the growing season; you can assume each crop will receive several in the spring and early summer. It shows, nevertheless, how preoccupied with chemicals farming now is, and how much residual material may at times be included in the air we all breathe.

The following programme might apply to a medium-sized arable farm growing wheat, peas, oil seed rape, sugar beet and potatoes. The wheat was sown in the previous autumn, and is immediately resown after harvest in August. Peas, oil seed rape and potatoes are sown in the spring and harvested in the summer. Sugar beet sown in the spring is harvested late in the autumn and stored until required by the beet processing plant.

The Spraying Year

	Chemical	Crop	Purpose
March	Dicamba or Dichloroprop or MCPA	wheat	as hormone weedkiller
	Metribuzin	potatoes	as hormone weedkiller
	Trietazine plus Simazine	peas	as weedkiller
	Chlormequat plus Carbendazym	wheat	as growth regulator plus fungicide
April	Phenmedipham	sugar beet	as contact weedkiller
	Chlormequat plus Carpendazym	wheat	"
	Phenmedipham	sugar beet	"

	Chemical	Crop	Purpose
May	Flamprop–Isopropyl	wheat	to control wild oats
	Mancozeb	potatoes	to control blight
	Triazophos	peas	as insecticide
	Pirimicarb	sugar beet	to control greenfly
	Triazophos	peas	" " "
	Pirimicarb	sugar beet	" " "
	Mancozeb	potatoes	" " "
	Triazophos	peas	" " "
	Trichlorophon	sugar beet	as insecticide
June	Pyrethroid	wheat	to control greenfly
	Mancozeb	potatoes	" " "
	Diquat	peas	to dry out the plants ready for harvest
	Triazophos	rape	after flowering, against greenfly
July	Diquat	rape	to dry out plants before harvest. Half the crop is treated this way
August	Glyphosate	wheat	within a few days of harvest to kill couch grass in stubble
	Organo-phosphorus	in grain store	to prevent insect damage to harvested grain
	Sulphuric Acid or Diquat	potato plants	to wilt them (as necessary) prior to potato harvest
September	Paraquat	wheat soil	to kill weeds
	Tecnazene	harvested potatoes	to prevent rot and sprouting
	Triadimercol plus Fuberidazole	dressing on seed wheat	to protect it from soil-borne diseases
October	Chlortoluron	wheat-sown land	persists in soil to kill weeds, before wheat-grass appears
	Propyzamide	newly-sown rape	after rape plants appear, as a weedkiller which may persist several months in the soil
	Pyrethroid	wheat	to control barley yellow dwarf virus, spread by greenfly

Chemical	Crop	Purpose
Fluthrin	rape	against stem weavil

November ⎫
December ⎪ everyone breathes freely, except game birds
January ⎪
February ⎭

We have, however, already seen the enthusiasm begin to fade from the protagonists of this approach. They are consciously on the defensive, unable any longer honestly to refute the indictments laid against them by conservationists and the public. The damage to landscape and wildlife is obvious and ugly. Farmers, previously confused and rather unwilling patrons of the industries which supply them, now find themselves heavily mortgaged and uncomfortably dependent upon government subsidy to keep their operations viable. This is a far cry from the self-sustaining, energy-efficient mixed farming pattern of former times, in which was vested the security of the entire community.

But the damage cannot so easily be undone. Farmers themselves would offer a stronger lead if they were less dependent on technical advice from their suppliers. For example, schemes introduced by sewerage authorities to give away clean, free sewerage sludge as fertilizer, were at first unwelcome to farmers – unaccustomed to thinking for themselves – since it meant the prompt withdrawal of advice from the fertilizer manufacturers. The agricultural support grants under the Common Agricultural Policy of the EEC, which tempt farmers away from good agriculture into high finance, need drastic but gradual revision. And the breeders of new plant species, already major contributors to the stock of crops able to cope with the chemical era, will have to develop a new generation of hybrids able to contribute to the revitalizing of the land. There are signs that the will is there, and in Part Three we show how to cooperate with and take part in the renaissance of health in soil and agriculture.

CHAPTER THREE
Tainted Springs

Of all the elements, water is by far the most vital. Oceans of it occupy the greater part of the surface of the earth, and have directly nurtured all life at all times from its beginning. Those forms which emerged from the oceans to evolve further on dry land brought with them personal reservoirs, and to this day the tissue fluid which circulates nourishment and cleansing power through the bodies of land creatures still in its composition recalls kinship with the sea.

Our weather begins with evaporation from the oceans, making possible the rainfall and springs of fresh water which sustain life in the soil. While human settlements were small or spacious, our ancestors could revere and trust the living streams which slaked their thirst, fed them and cleansed them body and soul. Water could be the stuff of myths and dreams, by turns cloaking deep mysteries, providing a frontier between life and death, and conveying heroic travellers on their journeys of adventure. It was the life-blood which flowed from their landscape into their veins and their culture, and back again. Water wedded them to nature.

We can understand why the engineers of great settlements chose watercourses by which to site their cities, and why they went to great lengths to carry water into their centres. It was not just a necessity, but a celebration. Great fortunes and creative energies have at times been applied to it, from Rome and Venice to Isfahan and Tenochtitlán.

Degraded by Abuse
Rarely was corresponding attention given to that more gutteral matter, the outward flow of water soiled by use. Until a hundred years ago only the Chinese had given it much thought; but they

excelled. For four thousand years they have collected human and animal night soil in covered vessels at the end of each city street, there to ferment to a safe odourless product which could travel by canal to the countryside and fertilize the land.

Elsewhere, inattention to excreta betrayed the unhealthy basis of the city's success. Crude open gutters might convey slops to the river stream, but most would lie and rot in soil and stagnant hollows around the dwellings which voided them. Pressure from a growing population soon overwhelmed these arrangements, and the stench of putrefaction betrayed accumulating reservoirs of the agents of disease. They got in the wells, infested insects and vermin, and became endemic in human beings. Centuries of fatalistic suffering were endured before lethal disease became connected with contaminated water.

We still do not properly understand, and we have never really faced, the whole truth. Life has been degraded in the industrial city; infectious disease is just one of its reflections, and the earliest to show. By failing to recycle our excreta efficiently to the soil, as the Chinese have always done, we have fundamentally defied the basis of our health. What should have composted wholesomely as food for successive crops of healthy plants has been allowed to poison us.

We did not lack the opportunity to appreciate our error. Albert Howard (pre-war pioneer of organic agricultural science) and McCarrison realized the truth early this century, and tried to persuade Edwardian England of it. They were ignored by all but a few, who between them have preserved their pioneer observations for the attention of a more receptive generation. By everyone else the pressures discussed in chapter two were more strongly felt; these colonial explorers were an embarrassment best forgotten.

An alternative answer, more consistent with the manufacturing and engineering age, was finding favour by 1851. That was the water closet, first installed for public use at the Crystal Palace Exhibition. It rapidly became fashionable, and the water carriage system of sewerage became almost universal. Great subterranean sewer networks were rapidly installed in all the major cities of Europe, banishing the problem from sight and smell but leaving it unsolved. For, even when sewage treatment works made sanitized sewage available, it could not be returned to the land because of the burden of poisonous chemicals it contained. No provision had been made to separate industrial effluent from domestic drains.

But water engineers can now take pride in drinking the liquid returned to the Thames after treatment at sewage works like those at Teddington. Except in times of flood, it is said to be fit to return to the mains. That is just as well, since every litre of water that flows down the river will be passed through the water systems of riverside settlements several times before being allowed to escape to the sea!

We have outgrown both our reverence for water and our traditional supplies. Thirsty cities have this century been forced to sink deep wells in order to supplement finite surface waters from sources underground. We are mining water just like any other material resource, refining it for a sanitized function in our technical scheme of things. The elemental bloodstream of nature, once honoured and celebrated in thoughtful cultures, has been tamed to the status of a sterile chemical servant in ours.

Problems with Nitrates

In chapter two we followed nitrate fertilizers and pesticide sprays as far as the soil, and watched the surplus flush through the land drains into ditches. The amount that goes this way varies, but is reckoned to include more than half of the nitrate applied in the first place. Sufficient pesticide accompanies it to alter radically the natural life of the watercourse; if a discarded container of pesticide concentrate has been thrown there, it can sterilize it completely. By a system of drainage channels appropriate to the landscape, rainwater from the fields is led to the streams and rivers, complete with its burden of nitrates and pesticide residues. Sooner or later, most of these residues will pass the intake pipes for a water treatment plant.

Nitrates could in principle be removed from drinking water, but only at additional expense. The World Health Organization and the EEC recommend that concentrations of nitrate above 50 parts per million should not be permitted in drinking water. That figure is exceeded regularly, sometimes doubled, in several parts of Britain.

The possibility that nitrates may indirectly cause stomach cancer has received a lot of attention recently, and remains a matter of controversy. But on one effect all authorities are agreed. Nitrates can alter and stabilize the chemical structure of haemoglobin, so that it is unable to bind and release oxygen – the essential property which enables us to refresh all our tissues from the air. This is the mechanism which makes respiration possible, which in turn is the foundation of day-to-day human metabolism.

The spoiled haemoglobin – methaemoglobin – is permanently useless and a dead weight, until destroyed and recycled as new. So exposure to nitrates can reduce the overall oxygen-carrying capacity of blood, and in turn weaken the power of all the processes dependent on it.

The only reason this is not a serious problem for most people is that nitrate generally gets no further than the liver, where it is metabolized to nitrite and other less harmful compounds. But since infants cannot do this until their livers mature, at a few months of age, they are very vulnerable to nitrate, and if exposed to concentrations above 50 parts per million can develop methaemoglobinaemia – a bloodstream seriously impaired by methaemoglobin. This is a real risk for any bottle- or spoon-fed baby; breast-fed infants are protected because their mother's metabolism prevents the nitrate from entering their milk.

Methaemoglobinaemia happens rarely in Britain in a severe enough form to be diagnosed by doctors, and it is not a condition which looms large in textbooks. Most doctors will be unfamiliar with exactly what to look for, and the advice we receive from government does not set out to remind us, or enhance our vigilance. It could easily be that minor degrees of the disease are passing unnoticed or being misinterpreted as something else. And some degree of impairment to blood efficiency is almost certainly affecting a proportion of babies, though not to a degree that would be obvious without laboratory tests.

This problem would not in itself cause allergy. But if it interferes even a little with tissue respiration, it must reduce our ability to detoxify chemical stresses and make us more susceptible to them. There is nowhere near enough evidence either way on this issue; we have yet to decide it is important enough to make the effort to find out.

We think it is important for two reasons. The children most at risk from methaemoglobin formation are also more vulnerable generally because they are bottle-fed; we shall explain more about why that is later. And the safe limit for nitrate in drinking water is easily surpassed by the concentrations typical of vegetables, on occasion by ten to fifty times! So even breast-fed infants being sensibly introduced to cooked fresh vegetable juice or purée may be affected, despite their maturity by this time, if the vegetable happens to be loaded.

Happily, the problem is not progressive and corrects itself, as the liver matures. It is certainly no reason to avoid vegetables: in almost every other way, they are important positive contributors to health. But it is sad to have their benefit spoilt by a misconceived farming procedure, and irritating that government experts arouse themselves so little to positive curiosity about new problems like this.

Reluctant to Act

The Joint Committee on Medical Aspects of Water Quality reported to its parent departments in April 1984 that its advice was unchanged, recommending that nitrates continue to be restricted to 100 parts per million in drinking water which has been the British limit for some years. They made no comment on the steady rise in nitrate concentrations in both surface and ground water sources. Nor did they mention the probability that ground water nitrates will go on rising now for at least twenty years, whatever changes we make in practice from now on. Water laden with surplus nitrate has been sinking into the subsoil for decades, each successive rinsing being more concentrated than the last. The oldest layers, contaminated perhaps twenty years ago, are now entering the aquifers. Those that follow will be progressively more heavily contaminated according to farming practice in more recent years: there is nothing that can stop them. What the Joint Committee will make of this remains to be seen: it will be tempted, we fear, simply to increase the safety levels!

It could, on the other hand, turn its attention to the further complications that arise in water treatment from the excessive presence of nitrates in land drains and streams. If pesticides do not kill them first, masses of tiny creatures thrive on this nitrate and choke the backwaters with weed and algae. The resulting biomass may choke itself to death, or be killed by a discharge of a pesticidal chemical after spraying nearby. It then rots into massive quantities of organic degradation products.

Among these is a wide range of hydrocarbons. When these present themselves at the chlorination plant of a water purification works, they react with the chlorine to form a range of volatile chlorinated hydrocarbons, like the carbon tetrachloride used to remove grease stains from clothing. These very irritant and unpalatable substances are giving water authorities problems, since no cheap method has yet been found for getting rid of them, even partially. It remains to be seen whether water technology can keep ahead of fertilizer use. At the moment, it is some way behind.

Nor does the Joint Committee seem to take seriously the problem of pesticide residues, which threaten us all insidiously and the hypersensitive quite drastically. Many of these residues, like the chlorinated hydrocarbons, can regularly be found in significant quantities in people with no occupational exposure to them. Some people carry around up to eight different kinds. They dissolve into fat and are concentrated there, building up appreciable reserves which flush into the circulation whenever fat is lost by weight reduction. All these substances are toxic, that is their function; sporadic bouts of unexplained serious ill health could easily be attributable to them. Yet no government adviser seems prepared to admit any concern that a problem may exist, or to see any point in monitoring the blood and urine of exposed animals and members of the public. Any doctor requiring these analyses in his patients is obliged at the moment to send the samples to the USA for testing. And plenty of them prove to be contaminated.

These agricultural residues are worrying enough, and we are saddled with them. Over other additions to our water we have in principle more control. Far the most controversial of these is fluoride.

Steel, Aluminium and Teeth

Putting fluoride in the water supply to stop teeth rotting is a very crude idea. At best it works a lot less well than good diet and careful oral hygiene, which we could very easily encourage dentists to teach their clients. We have no idea how much of it people are getting, and very little attempt is being made to find out. It begins to produce unwanted effects at only about three to five times the recommended treatment dose, which in the case of a drug would only be acceptable under much better controlled dosage conditions. As the stuff is coming at us from the air, from tea, from toothpaste and from food anyway, why do we need any more?

The truth seems so bald as to be ugly. We cannot get rid of enough of it in any safer way. If you find that as hard to accept as I first did, you will want to hear the story of how it came about.

Until the 1930s fluoride was regarded as a poison, and new boreholes containing any amount of it were considered unfit for consumption. Around those deep natural springs rich in it people showed mottling, disfigurement and brittleness of teeth, abnormal bone formation, and sometimes effects in other organs which were hard to attribute to a particular cause.

There is also mining of the deep deposits where it is usually found, such as fluorspar, apatite and cryolite. These ores are important in metal refining, steel production and aluminium smelting industries, which were beginning to take on their modern shape and size about this time. Unfortunately, the biggest problem was the disposal of the soluble fluoride wastes thereby generated. They could not be simply poured back down mineshafts, because they readily penetrate the potable ground water over a wide area and condemn it. They cannot be put in containers or ships and dumped at sea, because they rot their way out too easily. Only teflon, an expensive fluoride-containing plastic, will contain it.

By 1939 aluminium and steel companies in the USA were regularly being sued for dumping fluorides in local water courses, and producing unacceptable effects in human beings and wild life. Damages were running high, and in that year an aluminium company retained the services of an American chemist to help them find a solution to their waste disposal problem.

G. J. Cox had been studying dental caries since 1933 on a grant from the Sugar Institute Incorporated, Buhl Foundation. His brief was to find ways of reducing caries which did not involve eating less sugar. So his interest had been captured by evidence suggesting that a little fluoride helped to reduce tooth decay, in children at least, without producing disfigurement. His suggestion was in its own terms brilliant. Why not sell fluoride for consumption at the critical level which produces this effect, solving the problems of both the metal and sugar industries at a stroke, and declaring the solution a public health benefit?

There was a problem. The critical dose, one milligram per person per day, came nowhere near using up the fluoride fast enough. But putting it in the water supply at a concentration of 1.5 parts per million would give the same dosage, and use much more fluoride – most of the water is used for washing and sewage.

That meant a lot of changes, but they were made. From being considered toxic, fluoride was suddenly discovered to be essential for human health. This opinion is put forward in all modern texts on mineral nutrition I have ever seen, but none shows any evidence for it except these alleged dental benefits. Many authorities even admit that a fluoride-free diet is not lethal, as it strictly should be if it were truly essential for health.

Fluoride comes to us from the smoke of steel works, in fresh produce grown on chemically fertilized land and in processed vegetables washed in fluoridated water. Tea is a rich source, because of the soil it grows on. And toothpaste contains far more than 1.5 parts per million, since it is supposed to be swilled away; in fact young children often swallow it. In quantities this comes to one to two milligrams per day consumed in vegetables, and another milligram from every six cups of tea. Toothpaste can add another milligram if swallowed, and fluoridated water yet another.

There is no rational basis for fluoridation on health grounds – quite the reverse. It was known at the outset that, even on a one milligram dose per day, white mottling of the teeth would occur in ten per cent of the population. On four to five milligrams, it would be practically universal. We have encountered cases in children of vague general illness where reduction of the daily fluoride intake has coincided with dramatic improvement. Usually, avoidance of the toothpaste is enough in itself.

So the greatest marketing coup of the century was put through, with almost complete success. In those countries with significant steel or aluminium industries, it has been adopted widely. Professional associations of dentists and doctors have accepted the evidence presented to them by industry and taken part in local research studies to demonstrate the dental effect. They have in general been extraordinarily uncritical of this evidence, and are mostly unaware of the mass of scientific literature worldwide which contests its safety. This evidence suggests adverse effects ranging from interference with a wide range of metabolic enzymes to genetic damage, birth defects and cancer. None of this received serious consideration in the report 'Fluoride, Teeth and Health' produced by the Royal College of Physicians in 1976. On the other hand, the report appears to acknowledge that if we are to maintain these industries, fluoride must be safely disposed of and this is the most even-handed way yet devised. If we want aluminium and steel we must accept the penalty, or pay extra to recycle and ration what we already have. The best we can hope for is admission of the truth, and a transfer of the cost from public health authorities to the direct consumers of these metals.

Offensive Cleanliness

Most people dislike sanitized water principally because of the smell of chlorine. It is there as the cheapest, most convenient and most effective sterilizing agent we have. It is unfortunately also intensely irritant to some people, as we discovered about Paul (see page 8). Now let Anna's mother tell her tale.

Anna

'Anna has always been a sickly child, even as a baby, with stomach aches, respiratory problems, nausea, sickness and permanent sore throats. She was constantly taking antibiotics, which gave her sickness and diarrhoea.

'Her tonsils and adenoids were removed at the age of six, and her sinuses were washed out twice, but she still continued to be ill over the next couple of years, gradually getting worse and worse; and as parents, we didn't know what to do next.

'From age 8 I noticed whenever she went swimming with the school she came home very poorly with shortness of breath, nausea, streaming eyes, very achy and irritable; and it was suggested that this was because she didn't like swimming lessons.

'For over four years she was seen by many doctors and specialists at different hospitals as she was always ill. They couldn't find anything the matter with her, and they thought it was because she didn't like school. I knew myself this was all wrong as she is a very bright child, enjoyed going to school, and was never any problem to get back there after the few days she was too unfit to go.

'About eighteen months ago Anna was experiencing severe pain in all her joints which were also swelling up, especially her knees and fingers. She was examined by two rheumatologists, one being a very eminent specialist, who both said it was just a juvenile problem and dismissed anything we said. Their eventual answer to Anna's ill health was Gilbert's Syndrome [a disease caused by defective liver metabolism] but even this they didn't tell me. It was left to the general practitioner to tell me this, when I had taken her back yet again with severe pains in her knees.

'Fortunately, we took Anna for a second opinion to another general practitioner one day when she was very ill, and he suggested yet another rheumatologist. But this one thought Anna's trouble might be allergy-related, and suggested an allergy specialist. As a last resort we agreed to this, and it is as well we did.

'Anna is allergic to chemicals including chlorine, and had to give up swimming and even just being in the bathroom. If I am running her bath we make her leave, because it causes her to cough and feel very sick and dozy. She drinks filtered water, and has all her fruit and vegetables washed and cooked in filtered water.

'Anna is still receiving treatment for her chemical and food allergies, but since her treatment and testing started last August she is a completely different girl. Her health has improved dramatically and she can now see she will get better in the future.'

Chlorine is also corrosive and hard to match exactly to requirements. It has consequently taken a toll on the iron main pipes which convey it to our homes, and many urban water samples are thus a little rust-stained. Water engineers have lately introduced a chemical into badly affected supplies to mop it up. They have chosen the class known as polyphosphates. If you read food packet labels, this name will be familiar; it is used in foods to retain moisture during freezing, and to gel or stiffen processed meats. Polyphosphates will also chelate metals; that is, bind them firmly into the structure of the chemical. But this has drawbacks. Chelated metals are much more readily absorbed by the body than their simple compounds, and polyphosphates can pick up lead and copper from water pipes just as well as iron – especially in areas where the water is soft and acidic. This may result in appreciable increases in lead and copper absorption by the body, which contributes to the burden of mineral imbalances (see chapter six).

And it will surprise most people to discover that most pollutant asbestos is water-borne, as microscopic fibres of chrysotile. Anything up to a hundred million of these fibres have been found in a litre of water in some US cities, although usually there are fewer than a million per litre. The risk this carries is in the UK hard to quantify, as no corresponding data exists; in terms of death it must be very small. We know of no people intolerant of it in isolation.

Dangerously soft

Water arriving in the home may not yet have reached its final condition. If we pass it through a water softener, there should be a drinking water tap which by-passes the softener, since the salt residues in softened water are too high for health. Babies in particular should not be bottle-fed with milk feeds reconstituted with softened water. Quite a few softeners appear to have been installed without this by-pass – an omission which should be looked for and corrected.

Eventually the water reaches the hot tank, where it takes on appreciable amounts of copper; a prolonged soak in a hot bath can transfer part of this to the bather through his skin, contributing to a toxic burden and deranging the balance of several nutrient minerals. Polyphosphates are likely to enhance this effect.

We may draw water from the kitchen tap, and wash or cook with it. If we choose aluminium saucepans we introduce that metal into our food, again with the aid of any polyphosphate present. A wide range of toxic symptoms has been attributed to this in many individual cases. If we wash up in detergent this adheres to the crockery unless it is carefully rinsed. Its residues can be shown to damage the delicate absorbent lining of the small intestine.

So it looks very much as if abuses of soil and water correspond. One compounds the other, inescapably reinforcing, spreading and complicating each other's consequences. That is why we criticize such trends. They destabilize life at its base, making it progressively less secure and more frightening.

CHAPTER FOUR
Air No Longer Fresh

I live very near the vacant east coast of England between Grimsby and Skegness. It is a place of broad beaches, with the tide forever out; of dunes and sea birds, of small settlements and holiday camps. For a century it has been a convalescent resort for miners, and a watering-place for holiday-makers of ordinary means from the Midlands. It can still offer the traditional pleasures of the seaside, brought up to date by the micro-chip. But, like all such resorts, it owes its existence fundamentally to sand and sea air.

Whenever the wind is in the north-east we are ruefully reminded of that famous advertising slogan 'Skegness is so bracing'. Bracing it is, howling towards us across eight hundred miles of foaming sea. Despite exhaust from the oil rigs, it is still clean. And refreshed by long intimacy with splashing water, it has an electrical quality which benefits those of us who breathe it.

To step out of the train in London, Sheffield or Newcastle is to record a striking contrast. The impression soon fades, like all smells, but it clings to the clothing to be taken home. It is less on rainy days, a compensation for other inconveniences. During the still hot anticyclonic days of summer the smell is stifling.

Chronically polluted air, whether it occurs in our cities or in the country, should not be necessary and the human cost is high. If we have lost the smell of fresh air we should at least consider what is in the air we breathe and first we can begin at home.

Of Aerosols and Aeroplanes
Spray cans could never have caught on before those prosperous years of the mid-century. They are an expensive way of doing

many things we could manage much more cheaply in other ways. But consumer judgement is obviously not my longest suit. They are everywhere – furniture polish, air freshener, perfume, fly spray, window-cleaner, paint, deodorant, lubricant, adhesives, pretend snow, de-icer and medicines.

The materials which characterize the different cans vary almost infinitely. Inside it they are dissolved or dispersed in the liquid propellant. When the button is pressed a spray of fine droplets of the propellant squirts out, under the pressure of the gas in the can. Immediately it is free, the propellant liquid evaporates into the same gas, leaving fine particles of the suspended chemical dispersed in the air as a haze. Their fate after that depends on their nature, but most will stick eventually to any surface they encounter. If that happens to be the eyes, nose or skin of someone breathing the air containing them, there are all the conditions required for an irritable response. Now multiply by the number of puffs from any one spray-can, and the number of cans you keep, and you begin to see how a major stress has crept into the air of our homes. It should not surprise us that some of our children fail to tolerate it, and develop appreciable symptoms.

If you think that bad enough, consider the propellant. Without this, the spray-can is a non-starter. Whatever the material suspended in it, the propellant is roughly the same, one of a small class of gases called the halocarbons. These are derived from the methane which bubbles up from stagnant mud, by replacement of the hydrogen atoms with the much heavier chlorine or fluorine. So, although they are still gases in ordinary conditions of temperature and pressure, under moderate pressure they liquefy. Spray-cans are loaded with the liquid, and the gas trying to evaporate keeps up the pressure in it. So the very last trace of liquid squirts out as enthusiastically as the first, carrying its dose of the particular chemical which gives that can its function.

If you often reach for a spray-can for one purpose or another, you can appreciate how halocarbon vapour accumulates and hangs around in your home. Add to this other solvents used in cosmetics, adhesives, stain-removers, air fresheners and cleaning fluids. Then consider the chlorinated hydrocarbon insecticides used to preserve carpets and clothes; and the formaldehyde in proofed and treated textiles. Like the halocarbon propellants they dissolve into the air completely as gases, leaving no haze of particles to give them away. And some of them – acetone and ether, as well as halocarbons – are avidly soluble in fat and therefore capable of interacting much more intimately with the delicate surfaces of the eye, nose, airway and lung.

When present in even tiny quantities, like those in the immediate vicinity of the open bottle, they are obviously offensive and irritant to these membranes. And when they dissolve in the bloodstream they have some opportunity to dull our nervous systems centrally – ether, remember, is an important anaesthetic! The quantities required for that may be much higher than we expect in ordinary domestic circumstances, but these are high enough to raise doubts about their safety.

The breath of healthy people in ordinary urban surroundings contains a great number of organic solvents, measurable in parts per billion: workers in the relevant industries exhale a thousand times as much. There may easily be subtle and important consequences from such levels that we cannot yet detect. As we shall see in Part Two, there are plenty of puzzling conditions that may possibly be accounted for, if we set out to explore them with sufficient scientific curiosity. We lack not the lines of enquiry, but the will to pursue them – a conclusion we reach often in these chapters.

In the world at large solvents may have effects disproportionate to their quantity. With halocarbons this is already clear. There is no biological cycle for degrading them, so they accumulate in the atmosphere where they destabilize the thin layer of super-oxygen – ozone – which girdles the earth in the upper atmosphere. This ozone layer turns out to be a powerfully protective shield against nuclear irradiation from the sun – cosmic rays. These rays are already blamed for the skin cancer which affects some Europeans working in tropical climates, where our colouring offers inadequate protection against them. As more cosmic rays get through to the earth we can expect more people, and more forms of life, to be harmed increasingly.

High-flying aircraft and aerosol propellants share this unexpected power to touch global life at one of its weakest points. Maybe we need the jet age; that is debatable. But to hazard the atmosphere of your home and the prospects of your grand-children in frivolous puffs of chemical mist is indisputably perverse.

Smoke Screen

Tobacco smoking is even more self-destructive, but by now practically everybody knows it. There can be few who have never had acquaintances die or ail from its effects, and the propaganda has been pouring out from every direction for a long time now.

I always feel sorry for people who still smoke. They wear a hunted look, are much more apologetic about lighting up than they used to be, and find increasingly scarce the places they are allowed to smoke in peace. It must be very important for them to endure all that, and I would not willingly increase the burdens their habit is presumably helping them to bear. But I have to say a word on behalf of their children. They take for granted the smell of stale smoke everywhere, when they come home. And though they may not choose to smoke themselves, they are obliged to breathe the haze created by their smoking parents.

As active smoking dwindles to a minority habit under cultural attack, the predicament of passive smokers has gained in prominence. It has a large scientific literature all to itself. We know that passive smokers are more catarrhal, cough more, and get more ear infections and colds. Their colds are more often complicated with bronchitis or asthma, and their function is appreciably impaired. So is their resistance to other diseases such as meningitis. We must presume they accumulate cadmium and lead in their bodies faster than other people, because cigarettes are heavily laden with these toxic heavy metals.

These are just the irritant effects of the tiny solid particles of smoke. The combustion gases are more insidious. They still contain a little nicotine and cotinine, which we suspect make children who inhale them more easily fractious and irritable. Before birth, as the babies of smoking mothers, they grow less well; we have no reason to believe this retardation ceases altogether when they stop piping in fumes from mother's bloodstream, and start inhaling for themselves.

An Englishman's home may still be his castle, but it pays to bear in mind the prospects of his heirs and retainers. Young people forced to inhale the spent smoke of their parents are not only at an obvious and immediate disadvantage, but also put at risk the future function and health of their airways. And unfortunately, senses dulled by the habit of smoking are the last to realize the hazard they are creating for younger lungs and minds.

Heat Haze

Smoke from the domestic fireplace claimed a lot of post-war attention, and the Clean Air Act was an important success. Unfortunately, not all our heating problems were banished when fuel became smoke-free. The combustion gases are with us still. These dissolve into the air without a trace, like aerosol propellants. The vast bulk – water vapour and carbon dioxide – are harmless in themselves, though they dilute fresh air and reduce its net oxygen content. Water vapour is even helpful, moistening air excessively dried by central heating. However, not all paraffin stoves and gas cookers are efficient enough, nor sufficiently well ventilated, to ensure that there are no toxic by-products. The volume of output from these appliances is such that even tiny traces of such by-products can quickly add up.

The major hazard from fuel gas is carbon monoxide, formed when the combustion mixture contains too little air. Poorly trimmed wicks or smoky flames usually mean more of it. Its effect is deadly in any quantity, quietly locking up blood cell haemoglobin as carboxyhaemoglobin. Like the methaemoglobin (see chapter three), this is useless for carrying oxygen. So any carbon monoxide we inhale reduces the efficiency of our blood for respiration. The effect rapidly accumulates, and is now a common method of suicide. It just as easily kills accidentally, because we sense no danger or discomfort.

Non-lethal doses are therefore quite undetectable, and seriously impair our metabolic potential for weeks after just a brief exposure. There must at any time be thousands of affected children whose symptoms are taken for something else. This is easily done, since they so readily lead on to other ailments. Impaired respiration automatically reduces the capacity for every kind of metabolic effort, like maintaining immunity and stress resistance. So children affected by carbon monoxide are less tolerant of every kind of challenge. That some of them are easily infected and have multiple allergies, should come as no surprise.

Householders who have sought to reduce their heating bills by cavity wall insulation have sometimes been caught in another trap. One process in common use depends on a reaction between urea and formaldehyde, which forms an insulating foam. In some conditions this foam can release formaldehyde into the house, where it is very poisonous. A few people have actually had to move out of their homes, defeated by a chemical reaction the contractor is unable to stop. There are probably many more who are mildly affected, and have not yet appreciated the cause.

Improved safety standards have been recommended which should help to ensure that this process is not used in circumstances when 'gassing off' is likely to follow. We would still have misgivings about the use of foams in cavity walls, and feel happier about processes which inject solid material.

Acid Drops

Flue gases are not confined within the four walls of home. We share them in our cities, and in the world at large. The haze that hangs over every conurbation corrodes buildings, bridges and people. The acidic gases it contains irritate and damage our eyes and airways, and with smoke particles are the major life-long causes of destructive lung disease.

Even when swept up in our weather and blown away, these gases are fated to turn up like bad pennies – unacceptable currency in foreign lands. They dissolve in atmospheric water droplets, forming dilute solutions of strong acidity. Clouds of these droplets are borne by the trade winds of the hemisphere across national frontiers, eventually to fall as rain on hilly regions thousands of miles from their source.

Acid rain is the best-known European ecological problem of modern times. The Black Forest is one of many great woodland areas said to be near destruction from persistent rinsings. Since the forests provide nature's mechanism for converting back to oxygen the masses of carbon dioxide released in the combustion that makes the acid rain, we are at one stroke weakening and overwhelming this vital process.

Carbon dioxide is therefore accumulating in the atmosphere worldwide. This has disproportionate effects, since it lets in ultraviolet rays from the sun but will not let out the earth's infra-red rays. This is exactly the effect we use in greenhouses, to warm the soil and extend the growing season. A global greenhouse is gradually gathering in our atmosphere, and will eventually raise the average temperature everywhere. This may be attractive in a European winter, but think again. Melted ice from the polar regions would raise the level of the seas, flooding the lowlands where most of our best farms and largest settlements are. By burning fossil fuel as fast as we now do, we threaten at least our civilizations, and probably all life as we know it.

Exhausted

Smoke carries more immediate threats. Of the twenty-one most potent carcinogens known, eight are usually present in the smoke from fossil fuel combustion, and two more are likely to be formed in some circumstances. At present levels of energy use, smoke abatement will probably never reach the point where these problems can be solved.

To the sulphur dioxide, hydrogen fluoride, oxides of nitrogen and carbon monoxide contributed to the air of the city by combustion generally, motor exhausts add formaldehyde, phenol, a further range of polynuclear aromatic hydrocarbons and lead. Volatile lead compounds are a cheap means of improving the performance of petrol engines, but when inhaled by young children are readily absorbed; they impair present function and future potential in their brains and nervous systems. Industry and government have shown quite extraordinary unwillingness to admit this hazard. However, the UK Government has accepted the recommendation of the Royal Commission on Environmental Pollution contained in its ninth report. We have now reduced the lead content of motor spirit from 0.4 to 0.15 gms per litre, and are committed to phasing it out altogether by 1990.

To maintain the octane rating of four-star fuel, refiners have added an extra thirty per cent of volatile components, including benzene. The proportion in motor exhaust fumes is also likely to be high, since engines running on this fuel at high speed and load tend to degrade other ingredients to benzene as well. The overall output into urban air of polynuclear aromatic hydrocarbons, like benzene, is likely to increase – and these are one of the top twenty carcinogens mentioned already. Furthermore, experience elsewhere suggests that lead-free fuel may generate the conditions for greater ozone production. The persistent cloudiness of northern Europe would be all that is left to protect us from the high levels of irritation this substance causes in sunnier climates.

Perhaps all this will be resolved eventually, as engines are developed capable of operating economically on lower octane fuel. Any hidden costs will be borne by today's children, already breathing at a sensitive age what the air contains now. They are better off without the lead; but they are not spared entirely: tap-water remains a greater source of lead than petrol ever was, for inner city dwellers at least.

Motoring also provides our largest remaining airborne asbestos hazard: the dust worn from brake linings still in almost universal use, since adequate alternatives are more expensive. Yet we know very well the harm done to lung function and structure when asbestos is inhaled, to be retained in our lung tissue for the rest of our lives. Whether or not we get the kind of cancer asbestos can cause, we probably all lose some of our lung capacity this way.

Sprays that Stray

One form of pollution looms much larger in the country than in the town. Chemical spraying of the land, particularly from fixed-wing aircraft, has aroused a considerable and hostile public reaction in recent years, which is at last beginning to be taken seriously. Not only are hundreds of people getting sprayed by accident, and sometimes ailing badly, but there is also a less obvious result. After a crop is sprayed by air, it is easy to observe how part of the cloud of droplets drifts well away from the field intended as its target. On windy days, which do not always stop operations as they should, this drift can, to the onlooker, seem the major part of the result. These stray droplets dry out to become fine particles of dust, which can carry on the wind for miles. They are therefore available to be inhaled by summer visitors to the country, and to some extent by all of us.

This may account for some of the residual amounts of pesticides which can be demonstrated in the blood of animals and human beings for whom they were certainly not intended.

The Food and Environment Protection Act goes nowhere near far enough in attempting to control these pollutions. But reinforced by more sensitive behaviour on the part of farmers, very conspicuous recently along the East Coast of England, the benefits are worth having. Eventually more radical measures must follow. And, for as long as we continue to spray chemicals on the land, the pressure will be on to make this operation cleaner and more efficient.

Charged Atmospheres

The single most dramatic difference between the air in wide open seascapes and elsewhere is its electricity. Sea air is ionized into unstable particles with electric charge; there is not only a great deal of charge, but also a surplus of the negative type. The charge is brought about by friction with the earth, especially with moving water and is characteristic of mountainous areas too.

By nature we appreciate this electricity, and take it on ourselves when we inhale. It seems to account, at least in part, for the wholesome effect of visits to places with plenty of charge. It certainly helps to keep the air clean. Ionized radicals attach themselves to particles of dust which are then attracted to others of opposite charge. Coalesced clumps of particles eventually get too heavy to blow about, and fall out of the air. Even if we still inhale them, they are more easily trapped in our noses, so that less gets through to irritate our lungs.

This cleansing process neutralizes the charge, however, and often much faster than it accumulates naturally. Other features of urban life waste charge faster still, so that it never gets to our noses. Electronic appliances like strip lighting and television or computer screens attract and earth it away. Electric space heaters and air conditioners do the same.

So in air-conditioned work places where people sit in front of computer monitor screens all day, they become strangely confused and weary. This has been recognized in situations like airport control towers and ambulance control rooms. Where equipment has been installed to recharge the atmosphere artificially, the results have been gratifying.

Our domestic arrangements can be just as enervating. Fan heaters, televisions and strip lights abound here too, and sealed double glazing units cut down drastically the natural ventilation that would replenish it. Artificial pile carpets discharge air quickly, and dry dusters trap charge every time they are used. Consequently a sluggish and irritable child, camped for hours in front of a television, will brighten up if coaxed out to play in the open air.

So the countryside still has something going for it, in spite of chemical farms and tainted streams. But if we think denatured air is the only penalty of city life, we are wrong. For that is also where people are most separated from their natural food.

CHAPTER FIVE
Food in Poor Taste

Some time ago I met a farmer at a dinner party. He comes of a long line of modestly landed people, well-established and respected in the county. I don't doubt that he farms well, and makes the best he can of the resources at his disposal. I expect he is parsimonious with chemical inputs, despite pressures from salesmen. So his produce is probably very good value, by prevailing standards.

Even-tempered as he evidently was, something he had discovered about the fate of his potatoes had really upset him. He grows a few, and has never been tempted to make a serious business of it because he cannot rely on selling even the perfect ones for what they are worth. As it is, he gives to contractors for practically nothing the spoiled, deformed and undersized specimens rejected by his graders.

Not long before, he had discovered what happens to these rejects, by visiting the potato reprocessing plant to which they are taken. While they wait in the warehouse, rot sets into the scars; so the pile is treated chemically to deter this and control rats and mice. When their turn comes, the potatoes are washed, bleached, disinfected, powdered and dried. An emulsifier and a preservative are added, and the powder is packaged for sale as instant mashed potato.

His hosts at the plant had been quite proud of their accomplishment, and made no secret of the massive profit they were making. The farmer was first shocked and amazed, and then furious. Now he could see how the price of his honest whole potatoes remained so low. The public were making do with the rubbish he declined to offer for sale, and which they would not have accepted as a gift. Chemical processing, cosmetic packaging and beguiling advertisements were cheating farmers and public alike! I had not the heart to draw his attention to what manufacturers were doing to his other crops.

Once chemistry had created successful and prosperous industrial organizations with profits to invest, there was no reason why they should not diversify into food handling and processing. The early history of this development has several strands, each arising from a deeply grounded public demand.

Sweet Success

The first demand was for sugar. We can well imagine how that might have been popular. To begin with it was also scarce, and presumably expensive; so it was a prized article at the tables of more prosperous houses. The earliest records of sugar refining in this country go back four hundred years. By the mid-nineteenth century importers were industrializing their refinement processes, with Lyles of Greenock and Tates of Liverpool as the front runners. Their amalgamation created a near-national monopoly of sugar distribution.

But the enormous growth of demand this century gave rise sixty years ago to a new policy. Sugar beet began to be sown in East Anglia, and the British Sugar Corporation was born. Tate & Lyle helped BSC to establish the regional refineries which were quickly needed to cope with a rapidly increasing annual crop, and the British market has been managed between these two concerns ever since.

For about twenty years now the entire European market has been managed as a whole through EEC agencies. Quotas are assigned to member countries, and contracts distributed among farmers wishing to participate. Total acreage is controlled, intensive spraying programmes are maintained and delivery deadlines are set. At each area refinery, lorries roll in continuously throughout the processing season.

The end product represents no more than 16 per cent of the whole beet, and yields on consumption less than the energy it takes to produce it. White crystalline sugar is highly purified, at least 99.5 per cent dextrose. Very little is left to identify it as a food from a natural source, or to distinguish it from the products of chemical manufacturers. Indeed, in some of its variants such products are now legitimate ingredients. Icing sugar may contain up to 5 per cent of starch and 1.5 per cent of anti-caking agents, to modify its texture and behaviour in this use. And a kilogram of invert sugar syrup contains, as well as 500 grams of invert sugar, up to 15 milligrams (15 parts per million) of an antifoaming agent and 30 milligrams of sulphur dioxide as a preservative.

Fashionable Flour

Cooks found sugar a versatile substance, but to make with it the most delicate pastries and sweetmeats required a fine flour free from the coarse bran present in wholemeal. This had to be ordered specially from the nineteenth-century miller, whose raw stone-ground product would be riddled through a succession of sieves to provide it. The extra labour and time, priced for prosperous clientele, made white flour an expensive commodity up to a hundred years ago. And being a food beyond the means of the poor, it was aspired to as a status symbol.

Then engineers stepped in, with the invention of the steel roller mill. This worked much faster, was easier to maintain, and lent itself to industrial management. But best of all, it automatically separated the wholemeal grist into its separate components – mineral dust, bran, germ and the white starchy endosperm. Three of these had been prized special orders for stone mills; their ready accessibility in this new process immediately made it highly profitable. And the way was open to give the general public what they desired but did not deserve – creamy white flour, just like the gentry. By accepting it, at their own request amplified by marketing pressure, they devalued their nutrition by loss of minerals, vitamins, protein and fibre. The consequences are now well known, a tragic catalogue of degenerative conditions compiled from masses of observations on diseased lives, and at their expense.

The roller mill was the manufacturer's dream. He could continue to sell white flour at a premium, even after apparent concessions to his much larger clientele. And he had two marketable by-products: bran for animal feed, and wheat germ for the oil presses and feed compounders. His prosperity was assured, and the stone millers were doomed.

Centralization rapidly ensued, as in the sugar industry. In bulk storage, white flour proved less inclined than wholemeal to go mouldy, nourishing microbes as poorly as people. But long storage and market pressures created problems, nonetheless, which came to be solved with non-nutrient chemical additives.

So the ordinary white flour of modern times may in Britain contain ascorbic acid, potassium bromate, ammonium persulphate, potassium persulphate, monocalcium phosphate, *chlorine and *chlorine dioxide, L-cysteine hydrochloride, benzoyl peroxide, and azodicarbonamide. These serve as preservatives, bleaches, maturing and improving agents. One or two additional items are permitted in flour for special purposes, such as cake- or biscuit-making. And caramel may be used to turn white flour back to 'brown'!

So much is this flour devalued, and so dependent are we on it for nutrients not readily available to us elsewhere, that several are required by law to be put back. These include iron, vitamin B1 (thiamine), B2 (riboflavin), nicotinic acid or nicotinamide, and chalk for calcium.

If the flour is baked as bread, our staple foodstuff, it may in addition to traditional ingredients acquire acetic acid, monocalcium phosphate, acid sodium pyrophosphate, lactic acid, potassium acid tartrate, sodium diacetate and lecithin; plus various preservatives, emulsifiers, stabilizers, excipients and diluents from a permitted range. The grand total, for a typical white loaf, may exceed thirty items.

'Good for the Economy'

Throughout the long history which gave us chemicalized, standardized flour and sugar, one trend has led to another. The obvious need to remove dirt, and market pressures towards refinement of the whole foodstuffs, gave engineers and chemists their function. These invented and developed the necessary industrial plant and equipment, which required capital investment. That involved financiers, who strove to expand in search of the economies of scale. These were real enough to destabilize the early patchwork of many small operators; the large got larger at their expense. With civil servants, lawyers, accountants, financiers, engineers and chemists firmly in control the processing of these two foods became totally separate from their production.

The dangers of this separation are now obvious, but they were not in the beginning. The companies founded on it saw

* A voluntary ban on these two has now come into effect.

only a recipe for financial success, entirely consistent with trends elsewhere in industry. So it set the pattern for contemporary developments, and all the ramifications that were to follow.

Putting Food By

The big business takeover of food supply did not of course stop here. With canning, industrial methods began to supersede the housewife's crafts. Soldered iron cans lined with tin and shellac, manufactured automatically and filled on the production line, were the industrial engineer's answer to the housewife's glass jar. It rapidly took over, leaving only a minority of self-reliant enthusiasts sticking to the old domestic ways. There can be few houses left in Europe that do not keep a stock of tinned preserves or drinks.

Heat remains the mainstay of preservation in canning plants, but chemical antioxidants, preservatives and sequestrants have crept in increasingly as fine tuners of the result. Some of them get there from previous stages in the food handling process; some help to minimize the detrimental effect of canning itself. Even so, taints from contamination are often obvious enough to distinguish canned from bottled produce. A mass market tolerates this as the price of convenience, quickly forgetting the contrast.

Chemical preservatives really come into their own with more modern forms of packaging. Fast deep freezing became widely possible well within most of our lifetimes, but depends on techniques of preparation which much reduce the value of the product. In industrial freezing plants chlorine gas, volatile refrigerants, insecticides, texturing aids and colour preservatives provide mass answers to the processors' imperatives. But they make obvious problems for the process workers and hidden difficulties for some of their consumers. Even traces of gas and contaminants dissolved from polythene and aluminium packaging provide hazards for a few. Freeze drying overcomes some of these disadvantages, and enables company chefs to formulate from dried ingredients an entire recipe for reconstitution in boiling water. Successful ranges of quick meals in cardboard packets have been based on this technology, but they have never supplanted deep freeze packs and cans for staple items. Perhaps these instant foods are too expensive, perhaps not attractive enough. In any case, they often require emulsifiers, thickeners, anti-caking agents and stabilizers to ensure that the recipe reconstitutes smoothly.

Chemistry and preservative technology now seem inextricable. The manufacturer has set himself to get away from the drab appearance and taste that preservation necessarily entails. He cannot do it honestly, so depends on an illusion. The task is exacting. While the undertaker's embalming fluids need only save his clients' eyes and nostrils from offence, the food chemist's customers are expected to eat and enjoy the corpse as well!

Serving these purposes lies beyond the range of preservatives alone. The chemist must reinforce artificially the faded appeal of his product. His first step is to uncouple flavour from quality.

Isolating Flavour

It is important for us to appreciate the change this represents. Flavour is the natural counterpart of the supermarket bar code – the rectangular patch of dark stripes on white that the manager reads with a light pen to monitor his stock. The natural version provides us with instinctive and detailed information about what we are eating, and links our enjoyment of food directly to the needs it satisfies. Sometimes we are desperate for particular nourishment, and the food conveying it tastes particularly good. Other times, when we have enough, that same food interests us much less. We even know, almost to the mouthful, when what we have eaten just satisfies what we need.

Whole natural foodstuffs convey this information to us reliably. Even a toddler with no knowledge of nutritional science can choose from a range of plain whole foods set within his reach exactly the kinds and amounts which will meet his needs perfectly. If we marvel at today's electronic technology, this sophisticated natural mechanism should excite our admiration so much the more.

Herbs, salt, honey and even sugar can be used judiciously to improve the taste of traditionally preserved items to make their necessary consumption less unpleasant. These devices do not completely disguise the food's degeneracy, and we only eat the minimum we may require to survive hard times. When fresh alternatives again become available, we need no second bidding to revert to eating those.

But things have changed. Marketing researchers developing the interests of food manufacturers were quick to realize the commercial value of extending these principles. If sweetening and seasoning increases food consumption, let us have more of it. Sugar and salt quickly found places in a wider range of recipes where they had no legitimate function. From a seasonal exception they became a perennial rule.

That upset our self-preservative mechanisms. We regularly overconsume the sweetened or salted items, leaving less room for and interest in fresh produce. As well as unbalancing our nutrition, sugar and salt entice us further into dependence on packaged chemical food. We find ourselves buying them exclusively, consuming regular helpings of sweeteners and seasonings with every meal. Lacking the normal instinct to stop, we go on until we are gorged. Months of the habit add up to years, by which time we are grossly overweight, with deranged blood and metabolism. Degenerative diseases follow in a few decades, becoming commonplace at younger and younger ages.

None of this is the manufacturer's immediate concern. Inconclusive scientific evidence enables him to deny any responsibility. He can claim instead to be giving the public what we want, and in line with his policy of continuous improvement goes on looking for better artificial ways to enhance flavour.

The Taste Boom
One very successful line of research produced a whole range of intense artificial sweeteners which separate sweetness from calories. But their novelty presents problems. Saccharine, the earliest, is under serious suspicion of causing cancer in animals under experiment, and possibly in humans. At the least, it may make us more susceptible to the cancer-forming effects of other substances. In the USA it has been restricted in consequence; but not in Britain. And other sweeteners have followed it, equally novel and with controversial findings from safety research. We can never be completely at ease when safety testing is left in the hands of the promoters, the very people most interested in producing favourable results.

Another line of enquiry has proved equally profitable. The

usefulness of a traditional oriental fermented condiment was found to depend on its content of monosodium glutamate. Following precedents now well-established, this was isolated and manufactured for addition to a wide range of recipes whose own weak flavour could thereby be enhanced. Several other glutamates work in the same way. So do a number of salts derived from genetic material. All have entered the industrial chef's repertoire. Having the advantage of being derived from natural sources, they evaded the invidious 'artificial' category. Nevertheless, they share with other products refined out of their natural context a propensity for hazard. In recognition of this all these items are in Britain banned from foods intended for small children.

Even the successful development of flavour enhancers is a small triumph by comparison with the development of flavourings. There are thousands in use, usually in cocktails of several at a time. Many originate from natural material, not necessarily traditional foodstuffs; but we should treat with caution any suggestion that this necessarily makes them safe. Quite a large proportion are completely artificial. They are all kept hidden, even between manufacturers. Government permits them to be retained as trade secrets by their users, possesses no comprehensive list and makes no attempt to regulate them. Even current labelling regulations give no clear impression whether the flavour of an item is natural or artificially contrived. They seem designed to conceal and confuse.

Creating Images to Eat

I have in my hand a package, purchased recently from a local supermarket. It is a foil sachet, and contains a brand of strawberry flavour dessert mix. The contents can be mixed with cold water and will set firm in about ten minutes. The result is illustrated on the packet, garnished with a slice of strawberry.

Since there is no strawberry in the mixture, this narrowly escapes being in breach of the labelling regulations by describing the picture in small letters as a 'serving suggestion'. Indeed, there are very few ingredients that would be considered foodstuffs in the ordinary sense: the nearest are sugar, hydrogenated vegetable oil and whey powder. Of the fifteen specific

ingredients listed apart from flavourings, ten are there to colour, texturize, preserve and emulsify the other five. This is a
completely synthetic food, and it is not an isolated example. It is
among the more substantial items in the range that have come to
be known as junk foods. Their creators seem to subscribe to a
belief that since all food is in the end chemical anyway, why may
not the chemists invent a few of their own?

Very early in the development of this genre were several
economical imitations of expensive recipes. Ice cream and
custard powder are two examples which remain extremely
popular. By substituting cheaper ingredients for the cream in
one, and the eggs in the other, the manufacturers were able to
create cheaper but more profitable imitations, and enlarged the
respective markets enormously by sharing the extra profits with
their customers. The nutritive value of the products was considerably less, and they scarcely compare with their prototypes;
but as many of their consumers have never tried the real thing,
they cannot know the difference. These items have become for
them distinct foods in their own right: that they originated as
imitations ceases to matter at all.

It was from there a short step to synthesizing foods from
convenient ingredient substances. Considerable art and
marketing knowledge are required for this. Colours differ in
their inherent appeal, and must match the flavourings and
aromas chosen to blend with them. The mouth feel of the
finished product must be right. The package must be attractive,
with bright labelling and a natural appearance, and use crisp
images suggesting freshness. Advertising in the appropriate
media must portray exciting, amusing situations that are easy to
identify with or aspire to. In the food hall, packages should
occupy an impressive block of space and present a band of
choices; people then more easily forget the unstated choice, not
to buy at all.

Contesting the Doctrine

Some manufacturers may really believe in what they are doing
for the public; others pretend to. They all seem committed to
the prosperity of their industry. And they maintain a number of
doctrines to reconcile the two.

The first we have met already. It asserts that food is only

chemical anyway; we are simply joining nature in her own game. Many of the chemicals used in food are taken from natural sources; but some natural substances are quite poisonous. The public should not have such faith in nature in the first place. After all, nature provides the germs that rot and spoil food; customers would not be pleased if manufacturers allowed that to happen. People have a right to be protected from nature by vigilant chemists working on their behalf.

A second concerns cost. Cheap food is declared to be a right, demanded by the public. Manufacturers simply provide what customers demand. Without the benefits of efficient industrial processes food would cost more, and be inconsistent in quality. Variety in each season would be less.

A third concerns convenience. Busy modern housewives are supposed to need simple and quick meals to keep up the pace, and prefer to see the children eat up every scrap enthusiastically. A fourth concerns the safety of food additives. Each item is said to be very carefully researched before it is introduced, and there is no evidence of any danger from any of them in the amounts used in food.

What are we to make of assertions like this? They can be cunningly worded to sound like sweet reason. Some of the best image-makers available are at work on it. What is the truth?

By now, some 4,000 different chemicals are used in the manufacture of food, amounting to well in excess of 210,000 tonnes per annum. That equals at least 10 grams each per day, or 3.8 kilograms (over 8 lb) each per year. Dr Erik Millstone, Lecturer in Science Studies at the University of Sussex and a prominent independent researcher in this field, suspects the figure may be twice as much. That excludes sugar, which is officially regarded as a foodstuff. Biologically that notion is highly suspect. It may take its legitimate place as part of a whole food, and do no harm; but once refined out of such a context, it becomes aggressively deleterious to health at various levels. And on average we each now consume over a pound of sugar per week. Only forty per cent of that is bought pure, in sugar bags. The remainder is manufactured into the other goods we buy. To this collection we must add appreciable residues of the farm chemicals we considered in chapter two. In 1984 these were reported in a third of the fresh food samples tested by a group of Public Analysts, and there is no reason to hope this is not typical.

It is simply not true that none of these individual substances causes harm. The difficulty is in judging what results from animal experiments may mean for people. A Joint Expert Committee on food additives, appointed by the World Health Organization and the Food and Agriculture Organization of the United Nations, meets periodically to consider evidence put to them, and make recommendations. In many instances they suggest a maximum acceptable daily intake, which expresses a doubt about hazards at higher exposures; and in some of these instances their comments show clear dissatisfaction with the level or quality of information available. In Britain, reports from the Food Advisory Committee of the Ministry of Agriculture, Fisheries and Food reach similar conclusions. Seven years ago they gave manufacturers six years to produce extra evidence of the safety of some colourings used in food. At the time of writing, we still await their further report which was due for publication in 1985.

We have only to look up standard pharmacology reference texts to realize that to use some of these substances in foods is playing with fire. It is scarcely honest to state that no evidence of harm exists, when there is insufficient evidence of any kind available! Common sense alone forces us to the conclusion that continuous doses of a cocktail of 4,000 chemicals, amounting to half an ounce a day, must be doing at least some people appreciable harm. Perhaps, with the situation so evidently under-researched and unresearchable, the government is simply trying to keep us all calm. It would be acutely embarrassing for them to be forced to admit that a problem exists, because it would be impossible to remedy it fast enough to satisfy their critics. They do not effectively control the situation; no one does.

Meanwhile the domestic science curriculum in schools fails to encourage cookery pupils or teachers to exercise their critical faculties. We have heard of examinees penalized for avoiding convenience foods in their recipes. That is one area in which changes could and should be made, from the examining boards down. It is encouraging to see local associations of domestic science teachers beginning to take this problem seriously. They need parents' support.

But when all else has been considered, we come back to the manufacturers' most secure ground: we do, in fact, buy their products. We are taken in by the vivid green of processed peas, the beguiling simplicity of convenience recipes, and cheap cost rather than good nutritive value-for-money. When we exert different preferences, we can expect manufacturers to respond. There are signs already that this process is beginning, and some additive-free lines are being introduced in a few supermarket chains. Displays of bread in bakers' shops give wholemeal options much more prominence than they did five years ago, and specialist wholefood shops are thriving everywhere.

Unfortunately, we cannot even so expect a rapid and universal return to health. More is at stake than just the chemicals which manufacturers are putting into food. We have yet to appreciate the full implications of what they and chemical farming have taken out.

CHAPTER SIX
Too Much is not Enough

A few years ago we kept goats on our smallholding. We fed them from our own organic garden. We took great pride in the fact that, during the four years we kept four or five goats, none of the local vets ever had to call. And the secret of our success seemed to be our habit of always giving them plenty of fresh greenstuff every day.

Naturally they got the scrappy bits of herbage from the kitchen garden, and the rather tired outside leaves of their own crops until after Christmas, when we began to ration out the whole fodder plants. That meant that when in October local farmers were harvesting brussels sprouts and carting them off to the freezer packing plant, the stalks that fell off the carts on the bend near us looked very enticing by comparison with our goats' regular fare – big, rounded sprouts on long straight stems with not a blemish on them. So we offered them up, as a treat; which was dishonest, as we would not have touched them ourselves on principle.

But goats are sensible creatures: nor did they! While a scrap remained of our pathetic-looking organic garden produce, they ate this exclusively in preference to the visually enticing but chemically grown alternative. Only a day or two later, if we did not replenish their garden supplies, would they begin to accept the farm sprouts.

This made me realize the obvious: that to eat well, no special biochemical knowledge should be required. Nutritionists and dieticians are a recent invention, and we came out of some tough situations rather well in the thousands of years before.

Mankind is very old, and the landscape we grew up in is much older. Whether we consider the last few million years or the last hundred thousand, it is clear that great stability prevailed in our circumstances throughout our emergence as a species, and that all changes were gradual. It was in the lush tropical regions of the world that men and women first prospered well enough to leave traces of themselves. There would never have been a food problem in such places. An abundance of shoots and fruits must have been always readily to hand, besides game in plenty of a size for hunting once we developed to that stage. However, the present form of our intestines suggests very strongly that we have always fundamentally been much more dependent on vegetation.

These advantages had their drawbacks too. The vegetation was all but impenetrable, and insects as well as animals could turn the tables and live on people. These made time scarce for anything other than straight survival. The key to real cultural development lay much further north and south, in the more manageable regions where all our great civilizations were to thrive. Whether their founders migrated or were there already, they shared with tropical inhabitants a heavy reliance on fresh vegetation.

As civilizations progressed, organized crop and animal husbandry began. Whether as nomadic herdsmen or settled farmers, our ancestors gradually developed the skills required to secure food supplies. At this stage began the regular use of dairy produce and of grass seed varieties, the growing of which became more rewarding once prolific strains had been isolated and bred for agriculture.

So while our relationship with leaves, roots, shoots and fruit probably dates back millions of years, that with cereals and milk products may only span a few tens of thousands. And whenever we began to cook and eat flesh, digesting it involved a major metabolic effort. We can see how these more recent accomplishments are easier to upset than our ancient dependencies.

Even so, the instincts behind our senses of taste and smell have successfully grown to encompass these newer foods, because the pace of their introduction was gradual and there has been time enough. Up to the last century there was nothing we had altered so fast or so drastically that these instincts could be fooled or broken down.

These instincts enable you to read a loaf of fresh wholemeal bread like a book. Your appetite for it is keen in proportion to your need for its contents, and dwindles as that need is gradually met by successive mouthfuls. Chewing each one enables you to appreciate its quality and its quantity, holistically and almost instantaneously. Your interest in eating it reaches zero just as you exactly meet those of your needs wholemeal can provide.

The advent of food refinement has within a century thoroughly deranged all that. See what your instincts make of standard sliced white bread. It requires much less chewing, so you are tempted to swallow before savouring it thoroughly. The taste is different, lacks depth and fails to satisfy. So you go on eating. As you proceed your needs are met very unevenly, because contents always present for thousands of years have been removed. So you easily overeat the starch, long before getting anything like the vitamins, oil and protein your system expects. You never do reach a clear satisfaction point, but could go on eating until your stomach is too full to accept any more. By then your energy intake is way above your needs, and the seeds of obesity are sown.

All of this applies to your children even more than it applies to you. They cannot be expected to override with logic and self-discipline what their taste-buds tell them. The sensory basis of food appreciation cannot be easily restored. It requires the opportunity to savour good food again in favourable circumstances, and with plenty of practice. Most parents understandably give up the effort of converting children, long before good eating habits are established, under unfavourable pressure from friends and relatives.

Maladjusted Minerals
In chapter two we discussed how chemical treatment of the soil upsets its mineral economy. Our next line of reasoning can start from there. Considering how long ago chemists like Leibig began to analyse substances for their mineral content, it has taken us an age to discover how wide a range of them is necessary for healthy life.

We have known about the major minerals for quite a while – calcium, the chief mineral component of bone; phosphorus, sulphur and potassium, important components within every kind of cell in the body; and sodium and chlorine, the major mineral components of blood plasma and tissue fluids.

Magnesium and silicon occupy an intermediate position, required in appreciable but smaller quantities than these major minerals. It is about the minerals required in much tinier trace amounts that most of the recent additions to our knowledge have been made.

About iron we have known the longest, as the metal in haemoglobin which enables blood to convey adequate oxygen around the body. For this you only need about 5 grams of iron in your whole body, and as little as a ten-thousandth part of this is all you need to absorb daily from your food.

There are at least fourteen other trace minerals essential for human life, of which only zinc is required in quantities anywhere near iron – about half as much. And there are at least twenty others which occur in soils and in sea water but which are inessential for life or interfere with it.

All this is very easily thrown out of balance by neglect, and under pressure from conventional chemical fertilization. Even when the chemist realizes the complexity of the situation, he cannot necessarily preserve it once he is committed to artificial means. He puts himself in just the same position as a child who has taken a toy to pieces, and wants to make it work again. None of the mineral ores he can draw on as raw materials are entirely pure, and their trace contents may make his problem more difficult. Typically, far more fluoride and cadmium are included than the soil has formerly contained or can cope with; so these toxic elements are accumulating inexorably in chemicalized soils. On the other hand, insufficient zinc, manganese and chromium may be included to meet the needs of the plants and eventually of people, and superphosphates in fertilizer tend to bind up in inaccessible forms much of what there is. Any additions of these minerals may not be retained, and may even prompt soil stores to be released and leached away, having therefore the opposite effect to that intended!

Plants struggle to compensate for these inadequacies, substituting second-best minerals in their flesh where they can, but weakening as they do so. If an appreciable imbalance prevails in the soil, it will leave its mark inevitably in corresponding disturbances of the crops. Crops can obtain from the air and water the essential organic elements, nitrogen, oxygen, hydrogen and carbon, thereby adjusting soil deficiencies. But little or none of the other essential minerals are available that way. And though animals consuming deficient crops may in principle compensate for them by eating other herbage or licking mineral deposits, in practice under intensive rearing conditions their opportunities are limited to what the farmer provides.

We, the eventual consumers of both plant and animal food, spoil things further. The skins are specialized tissues, particularly rich in several nutrients including certain minerals. By refining this coarse material out of much of our cereal and sugar foods we drastically reduce their content of zinc, chromium and manganese. All of these turn out to be vital tools in our body machinery, and it looks as if many of us have lately become deficient in them.

Nitrates: an Embarrassment of Riches

Nitrates make people think of water, but that is only where the wastage goes. Let us now follow the fate of what stays in the soil and enters plants, because this turns out to be a greater cause for concern.

As for people, a little of what plants fancy does them good. Nitrates are also part of the plants' food chain in rich organic soil. In the natural order of things, nitrate salts are made available to the roots at about the rate that they can be used in the shoots to construct protein, and the amount to be found in transit through the plant stems is never large.

Even in an organic cropping system, plants can be put sufficiently off balance to increase the unused nitrate they contain. The most usual example is crops being grown well out of season, using glasshouses and heat to help them. This is stressful to the plant, which wants to be dormant then; it reacts sluggishly, like most people would if woken at an unearthly hour to work overtime. The effect is uneven: the protein-making process is worse affected, being complicated and more easily upset. Consequently a back-log builds up of nitrate absorbed more quickly by the roots, waiting in the stem to be converted into protein in the shoots and leaves.

In chemical systems this situation is chronic, whether under glass or in the open air. Nitrate is too easily come by, because it is applied to the soil already synthesized. The plant simply cannot digest it at the rate it can be absorbed, even if specially bred to grow faster. So masses build up in the edible parts, and are still there when the plant is consumed.

This affects vegetables especially, just the items we were beginning to think of as rather safe. From these we get over two thirds of the free nitrate we consume, if our experience is similar to the Swiss population studied by Professor Vogtmann a few years ago. The amounts are exceedingly variable according to the property of the plant and the season of its growth: in some samples of beetroot it can approach half a per cent of the weight, and winter lettuce has on average twice the amount contained by lettuce in summer. Overall, people are currently getting about a third of the maximum daily dose of nitrate considered safe by the World Health Organization. We can assume that vegetarians, or anyone eating a large proportion of vegetables whether fresh or preserved, are getting considerably more. Much nitrate-laden vegetation ends up as food for the animals fated to feed the rest of us, where we meet another whole range of problems.

Fast Flesh Food

Meat animals do not live long enough to get ill, but findings at slaughter suggest that by then they are heading that way. We even prize some of the signs. Take marbling, for instance. Streaks of fat through muscle make juices to baste it from within, so that it is more succulent when cooked. But in human muscle, marbling would be considered a degenerative disease. For it to be present in every child or adolescent would be disastrous. Yet we accept it in the animals we grow for food, and are not aroused to wonder whether anything similar may be happening to our own youngsters.

Fatty degeneration of the arteries certainly is occurring, and affects captive animals of all kinds as well as people. Fat consumption seems to be related, but not directly; the association is too weak to account on its own for much of the increase in heart disease. More plausible is the suggestion that the kind rather than the amount of fat is wrong.

We have already learned a lot by reflecting on the habits of our ancestors, many of whom thrived well on meat. But their animals were healthy and enjoyed a wide variety of food in all seasons. They ate grass not only young and green, but in seed a few months later. And they browsed leaves and shoots off trees as far up as they could reach. They still do, when they get the chance. But trees grow slowly, and pastures are usually only fenced off grass enclosures nowadays. No grass gets a chance to seed, and hay is deliberately cut before that stage. Seed cakes fed to animals have already had most of their oil removed.

So modern animals enjoy a much narrower variety of food than their ancestors, and far less of it is live when eaten. Whole seeds and dark green leaves are conspicuous absentees. In time past these would have been their principal sources of two special fatty acids which they cannot make from any other source: linoleic acid and linolenic acid. They are important, though only a minority item, because they enable the animal to fabricate special structural lipids which, in association with proteins, are essential components of the skins of cells everywhere in the body.

Without these the animals do not grow, so enough is added back to their diets to rectify this. Nevertheless, the proportion of structural lipids in the meat of farm animals still compares unfavourably with healthy wild animals and it seems probable that farm animals are relatively deficient.

Animal fats are now generally considered inadequate sources of structural fats in human diets, though this cannot always have been true. We are being encouraged instead to eat more of the vegetable oils rich in the essential fatty acids from which we can make our own structural fats. This may be insecure advice, as some of these – rape seed and safflower – are entirely new to us. But fresh whole wheat, dark green leaves such as cabbage, olives and nuts are safe sources, if only of small amounts.

Wherever they are to come from, we need them just as much as animals do. And we are learning how they are involved in other processes beside membrane construction. One is the formation of a whole range of prostaglandins, very short-lived hormone-like substances whose existence is quite a recent discovery. There are around thirty different kinds, some of which are involved closely in stress reactions, inflammatory processes, nervous function and resistance to disease. All these involvements strongly suggest a role in the phenomena of allergy intolerance and hyperactivity, and several parents do report their children much improved on supplements of essential fatty acids. The fat composition of modern diets is clearly quite important and may prove relevant to a whole range of conditions which puzzle us now, from allergy to multiple sclerosis.

This major issue is not the only problem meat presents. Animals not bred for health prove at times not healthy enough; so medication is often required. When regular antibiotic feeding was tried as a countermeasure, the animals fattened more quickly. So antibiotics are now routine animal food additives, as growth promoters. This worries many doctors, for various reasons. From our point of view, they could well be affecting the germs which live in our own intestines and condition their climate. If this is stressed unfavourably it spoils the efficiency of digestion, and may allow yeasts to preponderate dangerously since they are not affected by antibiotics. In certain conditions yeasts can transform from ordinary cells into a mycelium, a profusion of finely penetrating interconnected hairs capable of disrupting health quite disproportionately to their quantity. A few women nowadays certainly have stubborn recurrent problems with yeast infections. And some doctors are convinced that yeast mycelia are important contributors to food allergy and chemical intolerance.

Hormone implants are the other success story in fast meat production, despite the serious side-effects caused by one of the earliest, which is now withdrawn from use. The two now in favour are supposed to be used under veterinary control, but an active black market means they are available for irresponsible use. Residues found in a carcass would condemn it, but no mechanism exists to check every one.

Anything which disturbs the tuning of the body's hormonal system is likely to have widespread significant effects, and must be regarded with suspicion. Fortunately a recent EEC agreement will outlaw use of hormonal growth promoters within two years. Britain has opposed this move, and will be allowed an additional year. Whether the black market can then be effectively controlled remains to be seen.

Meanwhile meat has to be considered an expensive luxury which makes great demands of our digestive system. Its consumption in large quantities daily may continue to be a popular choice, but in our experience is a burden to the health of many who favour it. We should be exposed to much less immediate or potential hazard if it were grown slowly, for health rather than quantity. We could then, if we chose, safely enjoy the much smaller quantities we should be able to afford.

Eating Life

Cooking must, even through the coal age, have been a fearful nuisance. For meat eating it was essential, and in those houses able to afford it regularly a major allocation of space and servants went into preparing it. Only for baking, and tenderizing the tough roots of late winter, was cooking essential otherwise, and fresh herbs or fruits which were palatable raw probably got cooked much less or not at all. So throughout history until the present century quite a high proportion of most food prepared for human consumption must have been eaten raw or freshly cooked. Once refrigerators and deep freezers had been invented there was very little incentive left to eat food fresh out of the garden, even in households who were still growing it there.

Most people nowadays can only be getting really live food in the summer, so long as they favour salads at all. Frozen and canned foods have often been dead for months, though freshly thawed freezer food can look and taste much younger. By the time they reach the table all conserved foods have lost general nutrient value to a staggering extent, often in the case of many vitamins by more than half.

Obviously half the value of fresh peas in February is better than none at all; but if it replaces the genuinely fresh but drab food still available then, we lose out overall. Doctors are made very aware of the great susceptibility to disease and dismay that people show at this time of year. It corresponds with the hungry gap of former times, which ended in May with the earliest spring crops.

People wise enough to eat live food daily, summer and winter, are strikingly protected from this seasonal weakness. Why? The chemists protest that from their point of view, well-preserved foods and their living counterparts are practically the same. They therefore cannot account for the remarkable observations made by several biologists, the best documented of which are McCarrison's on rats and Pottenger's experiments with cats. These are long and detailed but for our purpose can be represented by one experiment of Pottenger's.

He fed different batches of cats in adjacent pens with contrasting food. One batch had raw fresh milk, the others had heat treated milk of various kinds. The raw milk batch grew and bred much better than the others, and their excreta proved superior as well. That result is preserved in photographs of their vacant pens, taken many months afterwards when beans had been planted in them. Luxuriant growth is shown in the soil fertilized by the raw-fed animals, in striking contrast to the mediocre results in all other pens. Even Pottenger could not fully explain this phenomenon, but he was convinced that live food conveys great nutritional advantage not explained by chemistry. It fell to a physicist of the same period, Simeon Kirlian, to make a remarkable discovery by which we can now begin to understand all this.

Kirlian found by accident that when exposed to a suitable high energy electromagnetic field, living things shine with a visible light that can be photographed. This Kirlian effect has remained highly controversial ever since, because it seems to add an extra dimension to living. So far, however, sceptics have failed to explain it away on a more prosaic level, and pragmatic seedsmen have got on with using it to predict fertility. Its potential for exploring human health has so far scarcely been touched.

If we take its findings at face value, they suggest that any living thing has a parcel of vital energy stored up alongside its chemical structure. It is very responsive to the varying fortunes of daily life, like fatigue, food, rest and exercise. Most people glow more in the morning, and are a bit dim after a hard day; you will recognize the feeling.

Plants grown in vigorous living soil have much more vital energy than others from chemical soils, a contrast which is passed on to the people eating them. And when they consume their food raw or freshly cooked, not frozen or preserved, the contrast is even stronger.

Using photographs of the Kirlian effect, we can trace the reason for this. When a food is first cooked its quota of vital energy immediately begins to leak away from it. Within a few hours it has gone totally 'dark'. If it is chopped up or chewed while still raw, the same thing happens. It seems that either way, when we eat truly fresh food (which still 'shines'), its vital energy is made available to us in a quiet explosion radiating from our mouths as we chew. But if we cook it first and delay the meal, leakage diminishes our portion. This probably unlocks the mysteries of taste, which we marvelled at earlier. It certainly fits many everyday experiences. Cooked food kept overnight never has such fulsome flavour as it did just out of the oven.

That should make us much more wary than we are of the freezer, cans and packets. Living food protects, conveys vigour and secures the same for future generations. Anything else allows these vital factors to run down.

The Undernourished Self

McCarrison, Pottenger and all the other biologists who have ever studied health, have celebrated the strikingly positive, whole and definite impression a healthy creature makes. Everything about his being is harmonious and fluent, but far from blending mildly with his surroundings he stands out within them, clear-eyed and confident. It is quite impossible to ignore his presence in the scene.

All the great heroes and heroines of legend and saga had this quality, and by its power lifted their followers above themselves, prepared to attempt great and dangerous adventures with confidence, proof against any assault. The ageless appeal of such heroic tales lies in their resonance with life as it bears on each of us. They uphold the vision of the strong and healthy being we each can be, master of the space granted by nature, holding all neighbours in high mutual regard, and trading peaceably with them for every need. There is no question here of assault by one upon another, nor any place for aggressive self-defence. Such things arise only in houses divided, where counterpoise and grace have given way to debased values and ugly behaviour.

Intolerance and allergy are of this kind, and have this fundamental origin. In a world degraded of its finer points by brutish exercises of material force, we have lost the secure integrity of healthy beings. The clumsy gestures our bodies make to preserve the health remaining are the exaggerated self-defence of organisms in disarray, imperfectly nourished from and threatened by surroundings themselves under corresponding pressure. Sometimes we appear too ready to succumb, sometimes too violent to oppose the jostlings and encroachments such situations inevitably entail.

Violent intolerance is a very conspicuous sign of trouble, and it is everywhere. We see it alarmingly often in the bodies of our children, and call it allergy. Intolerance erupts on picket lines, in football grounds, and on once peaceful streets. We mentally place this in a different category, but it is all the same. Alexander Schauss, an American scientist in the probation field, has collected convincing evidence which is now receiving serious consideration by penal reformers on both sides of the Atlantic, that much criminality and social violence can be related to unbalanced and inadequate nourishment. Astonishing results have been achieved by simply feeding prisoners better.

Meanwhile the quieter and less obvious signs of weakened self-maintenance are almost universal now, and it is frightening how little we seem to mind. Our bodies no longer resist infection or degeneration well, and we simply reach for medicines without any concern at our increasing dependence on them. Even when well we endure hardship poorly, our efforts lack stamina, and we wilt easily under moral pressure.

None of these failings is fundamental in itself. Underlying them is loss of a positive attribute, and that is what beggars us. We can no longer muster the vigour, resilience, stature, grace, mental clarity and unity of purpose which mark out health. If we may speak of rights at all, ours are these heroes' virtues. In seeking to supersede nature we undermine the abundant living of whole beings. That is the price we pay, individually and as nations, for dominion over a material kingdom.

CHAPTER SEVEN
Ideologies at Odds

Vance Packard was a name to be conjured with around 1960. His book *The Hidden Persuaders* was required reading, and opened the eyes of many young people like me, just forming their attitudes to the real world. Advertising was already an old and thriving industry, but it had not long formed the alliance with psychology which was to transform its impact to the powerful punch it commands today. Packard introduced us to the principles at stake, and gave new meaning to the age-old warning: 'Buyer beware!' His revelations must have saved me thousands of pounds over the years, though I have still occasionally fallen victim to a few of the enticements.

I had actually met marketing face to face, a few years before that. My father worked in commercial management for a multinational producer of food and toiletries, and it puzzled me that they made so many different kinds of toothpaste. Why did they not just pick the best of each kind, and put all their efforts into selling those? He explained. If a housewife goes into the chemist and looks for toothpaste, she will find all the different brands set out in little piles next to each other, on the same stand. She has no time to compare them methodically, and usually chooses what she had last time by the colour of the packet and its big bright lettering – or even by knowing its usual place on the shelf.

But perhaps the family has got bored with the flavour, and she wants to make a change. This is an important moment for her previous brand's rivals: which of them will she choose? A familiar name must jump out at her from a recent advertisement, but she can easily miss it as she scans the shelves. To catch her, a hopeful manufacturer must occupy as many slots on the stand as possible. If he has interests in five out of fifteen options, he stands much more chance of making his sale than if he only makes one out of ten. Multi-company corporations did not get big just by bothering – they used tactics like this to increase their market share.

You may think that once all the small toothpaste brands had been bought into one or other of the rival corporations, the incentive to operate this way would cease. Up to a point that happened, and a few brand names disappeared. But there remain many advantages to keeping one's eggs in several different baskets; the basic pattern still holds good. And it is not confined to toiletries. Enter a supermarket food hall, and you face the same phenomenon. Large arrays of tins, packets and bottles in each section appear to offer you endless choice. Prices and sizes vary according to inscrutable formulae which defy mental comparison. Brands glare at each other from adjacent stacks. Opt for one, and you still need to choose from a range of flavours. There are decisions, decisions, all the way.

Make no mistake. All this apparent choice is an elaborate, deliberate illusion. None of them is more than fine tuning within a very narrow band. If you want toothpaste, nearly all the brands have fluoride and sweetening: old-fashioned denti-frice is scarcely available. If you want baked beans, they will all be of one type, and almost all contain an appreciable quantity of sugar. Processed green peas will all be coloured, usually with the same combination of artificial dyes. You can choose between rival images, but the basic commodities they portray are mostly very much the same.

Behind the arrays of contrasting packages is concealed an opposite trend: steady shrinkage in the range of our real choices. The manufacturers have already chosen for us. Careful review of the performance of each line under their control has led to gradual pruning away of the least profitable, consolidat-ing the remainder on the smallest variety of fundamentally distinct food commodities and processing methods that is compatible with a comprehensive product range. This lowest common denominator of technological provision underlies the industrial efficiency which defeats the corner grocer, and maximizes supermarket profits.

From Mixed Farm to Monoculture

What has happened in the shops corresponds exactly to trends on the farm. Precisely the same logic applies, developed under the careful orchestration of the same corporate multinational boards of directors. Within each organization the cogs of production and distribution mesh smoothly together in a world-wide machine.

Wherever in the world the landscape permits, fields are getting larger. This cuts down on non-agricultural dead space, justifies larger machinery, and increases productivity per employee. Huge capital is tied up in machinery investment, exerting pressure on farmers to specialize. Traditional crop rotation patterns frustrated this development until systems of fertilization and pesticide management were devised which enabled similar crops to be grown for several seasons in succession. This maximized the efficiency of capital deployment in the industry, and relegated crop rotations to a secondary status in agricultural technology – one of many such steps away from naturalistic principles.

To meet the shifts in demand created by distributors, new crops are invented; we met sugar beet in chapter five. Oil seed rape is the most spectacular recently, part of the development of vegetable oil margarines in competition with butter. Here was a potential, recognized from the success of much cheaper early butter substitutes, and translated into practice by a major postwar movement of capital and effort. The interlocking redeployments these involved, of large sectors of agriculture and industry over a span of three decades, illustrates what power the architects of such campaigns can bring to bear – quite large enough to influence the economic policies of whole nations. How else would we come to eat potatoes from Egypt, or strawberries from the fringes of Sahara? By encouraging the specialized production of a few key crops in each region of the world and trading these globally, the multinational business corporations have realized an ideological dream. They have developed the chemical and engineering principles of the materialist age and imposed them on the living world. From pole to pole, the soil and all its industrially managed products express a totally mechanical-chemical world view.

Mineral Machine, or Global Organism?

We do not happen to believe in the chemical engineer's vision of universal order. But to criticize it we must state the basis on which we stand. Biology goes beyond chemistry, and operates by principles more subtle and apparently more conscious. Nature behaves as a whole, as if exercising her mind. For thousands of years mankind has had the notion that the earth as a whole is a global organism, a mother-god – Gaia. That idea has been expressed in ancient and widely differing cultures with remarkable consistency. And recently ecologists have been working back from modern evidence to this same ancient conclusion.

If this is anywhere near the truth, then it gives us a very different view of the chemist's impositions on the world. The chemist defies natural principles, and has supplanted them by his own. If we are just one tissue in the whole body of Gaia, it is as if that tissue has rebelled. And that rebellion not only disorders all the other tissues but also disfigures us, the cells in the rebellious tissue. The situation is indistinguishable from a cancerous growth in the global organism, originating in the tissue which is mankind!

The alternative conclusions suggested by these two different points of view could hardly contrast more strongly. Are we the masters of the world, in command through science and technology of its resources? Or are we a malignancy, disfiguring ourselves and it?

Consider the evidence. Our masterly manipulation of world food supply has clearly not yet benefited everyone, 'Liveaid' notwithstanding. To be fair, let us confine our attention to those populations in Europe and America who have benefited most from it so far. We have at our disposal plentiful fresh supplies of dairy products, vegetables, fruit, meat, eggs and cereals. Anything we desire is available to us, at any time of year. We ought to be the healthiest people on earth, not just the wealthiest.

There is no impressive evidence to support this. Most of us now survive to forty. Length of life after that does not appear to be increasing, and health sufficient to enjoy advanced age actively is conspicuously lacking in old people's homes. Serious degenerative disease of the heart and arteries prevails from a very early age, killing a regular trickle of people in their thirties. Cancer seems inexorable.

Sins of the Fathers

But the most alarming evidence is seen in the very young. Surprising numbers of them fail to tolerate their food, suffering instead a wide range of distressing symptoms. It is not just the newly concocted 'junk food' items, nor only the additives and pesticide residues. Some children cannot seem to tolerate fresh milk, however cleanly it is produced; or wheat, even grown organically; or eggs, whether free range or not. Exposure prompts in them signs of irritation and distress of one or several systems of the body, including the nervous system.

The effects are strikingly reminiscent of situations visible elsewhere. Worms and insects also react adversely in soil dressed with mineral fertilizers. They appear not to like it, yet it is not clear why. To account for food intolerance is even harder. Perhaps incessant exposure to a fairly limited range of abundant foodstuffs, without any rest between seasons, is a 'stress' in Selye's sense. Our particular liking for the complex foods incorporated in our repertoire more recently (see pages 62–3), would certainly put us under pressure. Perhaps all food is antagonistic really, but we formerly concealed the fact by eating a wide variety of different things in short seasons, just as food-intolerant children are now obliged to do.

There seems little doubt that this problem is now slowly being recognized. It is hard to believe it represents merely a shift in medical fashion. Allergy and intolerance of all kinds amounted to only a few per cent of childhood illnesses, perhaps as recently as twenty years ago. Now it seems to account for at least 15 per cent, maybe twice that much; and in most cases the provocative factors are mysterious. Every clinic, however successful they may be at analysing causes in difficult cases, admits to a proportion that baffle them completely.

The scale of the problem is one of the reasons why the majority of the medical profession still refuses to believe in it. That the bodies of crowds of us have just gone feeble all of a sudden, is inconceivable. And the alternative idea, that we are making the world a much more hostile place, directly opposes contemporary faith in science and technology.

But to us it is the only notion that rings true. If Gaia were threatened by a malignant tissue, would we not expect her to react just as we do in our own bodies? We would inflame in the face of such a development, mobilizing fierce resistance to its further encroachment upon our vital processes and well-being. To the fresh young cells of the malignant tissue, unaware of the outrage their culture has provoked in realms far beyond their ken, this resistance would appear remarkably hostile and unaccountable. It could well damage or stress their own vital processes, just as it is intended to. Until we actually gave up the struggle to survive, this battle would go on between our defences and the malignant tissue, which would never 'understand' what it was all about.

The analogy is strikingly close to what humanity is experiencing, at the vanguard of the technological revolution. If we are alien to Gaia, then she turns on us some of the tremendous resistance she is capable of. And if that notion sounds fanciful, how else do we explain the unremitting suffering large numbers of our children seem obliged to endure?

PART TWO

THE CHILDREN

CHAPTER EIGHT

Heavy Breathing

Your first nine months of life are spent in the certain knowledge that you are all there is. No alternative can possibly occur to you. Your universe is an intimate sea of warmth, movement, sound and occasional dim light; quite indistinguishable from yourself. There is sometimes pain, apprehension, panic, noise or violent movement. But mostly you are content – being, floating, unfolding, becoming aware; throbbing to the rhythm of the two hearts at your disposal.

Then comes the day when you break through the surface of that universe, to be born into another. If you are lucky your first moments are supported by your umbilical lifeline, and you have time to sip carefully at this new thing, air. More likely its coldness and dryness make you gasp. And from that moment hundreds of other micro-organisms, competing for life, enter your nose and mouth with that first rush of air, some passing deep into your passages and lungs. They settle there, stuck to the wetness of your skin, and try to carry on their lives. You begin learning to defend the space granted you by nature from being overrun. Until you die, you must not fail in that.

So you start by sneezing and coughing, to expel forcibly the intruders. You make an antiseptic mucus, sometimes policed with white blood cells, to engulf and destroy them; that mucus too must be expelled eventually. Mother feeds you her milk, with antibodies you can use until you learn to make your own. The first part of that lesson will take you many months – almost as long again as you have lived already.

Airborne Skirmishes

It is easy to see why the skin of the nose and breathing passages is so reactive. It was the first tissue to be active in self-defence, and throughout your life it copes with far the largest turnover of material to enter and leave your body. That material is dry and cold, and must be conditioned to the warmth and moisture of your lungs to ensure that no damage is done to them. Dust must be filtered out onto the lattice of hairs in each nostril, and smelt warning of poisonous fumes must activate the expulsive efforts, gagging and bronchial spasm, which bar entry to that most delicate and vulnerable lung tissue.

So brief running of the nose, sneezing or coughing are evidence of a nose at work, not really diseases in themselves. Even that is uncommon in most people, provided the air is fairly clean, not very dry, nor extremely hot or cold. Only when the defences come under serious challenge does anything really happen.

Suppose one day a colony of a virulent cold virus settles on the skin of your nose. It sets about digging in, and will attempt to colonize the cells immediately around. If you are not well defended, it will multiply, break out a few days later and make for the circulation.

You anticipate this and release a little histamine, your secret weapon. That opens up the circulation, fluid swells into the area, and the mucus cells go to work. White blood cells roam around, looking for virus particles and destroying them. For a few hours it is mayhem, and small wonder that you can feel a sore patch in there. But your defences are pretty tight, and you notice no more.

Caught on an off-day, things may go further. A few virus particles find homes and incubate successfully. A day or two later they burst out in their thousands, leaving the wreckage of these cells behind. This time you are overwhelmed by numbers, and the sore patch spreads. Skin all around the area becomes inflamed – more histamine, more circulation. It swells and thickens, and mucus pours. You probably know the feeling.

If your luck is out the invader makes it through your skin altogether, and the fight is carried to the tissues underneath. There thousands of fine drains lie for plasma spillages to find their way back into the blood. They are filtered repeatedly on the way through pellets of a glandular tissue, the lymph nodes. There are great pads of them around the nose and throat – the adenoids and tonsils. When the viruses get in there they are trapped, and the riot breaks out afresh. The node swells and hurts, but eventually settles to the steady business of preparing a permanent antidote to this particular brand of mischief. The buck stops there.

So a typical cold or sore throat scarcely breaks the surface, a skirmish at your main defences. Its symptoms are evidence that you are having to arouse yourself to beat the intruder off. In health you will rarely even experience this. Viruses will frequently land on your nasal mucosa and be efficiently and economically dealt with, without so much as an itch to show for it.

Not Quite as Planned

During the first six months of life while you are learning all this, it is important to be properly equipped and meet all the lessons in the right order. If your mother is unwell, poorly nourished or chemically polluted, this gets in your way. Your first breath enters a body imperfectly built and ill prepared for it.

So instead of settling down comfortably in your new surroundings, you make mistakes. Your tendency is to overdo things: too much histamine, too much snuffly blockage of the nose, too much mucus. You can scarcely afford to under-react, or you may not survive for another chance.

In infancy you do not yet really know what you are reacting to. You are newly aware of regular challenges to your breathing passages from the outside; but there were always many challenges from within you as well. When your body hurt or felt anxious before you were born, you could not focus on the problem and there was nothing much you could do about it anyway. Now you are born, the trouble always seems to be in your nose. And the great thing about trouble is to fight it off.

So you may well have reached the preliminary conclusion that, if there is trouble anywhere, you must respond in your nose. If mother is polluted and her milk antagonizes you, snuffle. If she feeds you with some other milk you can scarcely make sense of, snuffle harder.

That seems to be how Jeremy saw things. He put a lot of effort into his colds as well. But it turned out he was capable of much less clumsy responses once his immune system got its act together.

Jeremy

10 September

'Jeremy has been allergic since shortly after birth when he vomited a lot and had a snuffly nose. His mother cut out cow's milk and eggs and he was a lot better. He had a great many breathing problems from time to time with recurrent bouts of asthma and croup, and when he has colds he is extremely unwell. He has bouts of diarrhoea whenever he eats greens, eggs or milk; also soya and other pulses. His abdomen is frequently distended and his stools are fatty. His mother has tried to re-introduce foods, but has not been successful. She has cut out grains, additives, beef, pork, eggs, milk and pulses from his diet.'

He was tested and given immunotherapy.

14 January

'Jeremy's diarrhoea and vomiting have settled on his treatment with immunotherapy, but he still continues to have some respiratory tract infections.'

He was given treatment to boost his immune system.

26 October

'Jeremy is now extremely well. He is only unwell on very occasional days. He takes his antigen therapy both to prevent and treat off days.

'He will have treatment to boost his immune system if he has much respiratory infection this winter.'

Over-reactions take other forms just as easily. Neither Ian nor his parents had any means of telling what he was reacting to, so when he always came out with a full-blown upper respiratory response they naturally assumed he was getting repeated full-blown infections. Some of them were.

Ian, aged 12 years

25 April

'As a child he was always subject to tonsillitis, croup and infections. Practically every month he would have an infection and he recurrently had antibiotics for these. Between the ages of 8 and 9 he was considerably improved and then developed malaise and infections once more. He had shingles when he was 9, with a rash around the left side of his trunk. Since then he has spent much of his time at home in bed.

'He has vomiting episodes, when he vomits eight or nine times during the night, has a high fever and gradually improves over the course of the next six weeks. He feels ill during this time with malaise and anorexia. He suffers aching muscles and cannot undertake more than half a day at school at any time. He has also complained of a lot of abdominal pain independently of the vomiting episodes.

'His diet is unusual for a child in that he has a passion for cockles, mussels and prawns, and will have these every day. On the other hand, he dislikes milk though he is encouraged to have it. He had severe infantile colic as a baby and stopped breathing on several occasions while he was being artificially fed. In addition, his mother was on a milk-free diet and an egg-free diet for abdominal pain as a child, though she has these items now, but does not choose to have very much of them.

'Ian is very thin and pale. There are small lymph glands enlarged in his neck.'

He was treated with a milk-free diet initially, then on a varied diet. He had immunotherapy and some vitamin and mineral supplements over the next year.

11 months later . . .

'Ian has improved since starting on immunotherapy. His mother says that prior to treatment he could only attend school on a half-day basis, but now he is able to attend for four full days and a half day. He is generally feeling much better and able to cope with school work, and he has been working very hard. His appetite has improved and he is starting to put on weight.'

Ian proved to be sensitive to some chemicals as well as foods. His illness was triggered by a virus, but after that his vulnerability kept it going.

10 months later . . .

'Ian continues extremely well. He needs to have re-testing for his vaccines once every twelve weeks or so and I think this may reflect changing seasons and also the fact that Ian is growing a good deal.

'He is not able to do games at school and may try to do gym or yoga, for some physical exercise. He is also going to try Evening Primrose Oil as a supplement.'

Notice how long this went on, ruining half his childhood. This is not fanciful or far-fetched. If an infant cannot quickly reach the right conclusions about self-defence, he cannot build more advanced responses to them. His whole programme of immunological development, which is set to carry on regardless in pace with his growth, breaks down and stagnates. Immune deficiencies, whether medically detectable or not, are bound to follow.

Ears are commonly involved in the catarrhal reaction, because they are connected inwardly to the nose cavity. When the nasal skin is continually swollen the lining of these connecting drains swells too, blocking traffic both ways. If the ear cavity is dry, the air in it dissolves into the circulation leaving a partial vacuum which sucks the eardrum down onto the back wall of the ear, like vacuum-pack polythene. Or if catarrh is formed in the ear out of sympathy with the nose, this fills the cavity and cannot get out. So the eardrum bulges in the opposite direction. Either way it hurts, and hearing is impaired. Sometimes the trapped catarrh becomes infected, and the child gets very ill. Or the eardrum bursts letting the catarrh out, which is a natural 'grommet' operation. This only leads to a permanent hole in the eardrum if the catarrh does not quickly dry up and allow it to heal.

This was one of the features of Ruth's story.

Ruth

8 May
'Ruth has had eczema since birth but her main problems lately have been catarrh resulting in deafness.

'She has been on diets eliminating a number of common foods to try to minimize her eczema and her mother had suspected milk products and chocolate as provocants. However, eighteen months ago and again last year at Christmas, dried fruits and mince pies provoked severe behavioural problems – lack of cooperation, and tantrums.

'There is dry scaly eczema of Ruth's skin generally, but her ears are all right at present. She did, however, vary in her attention to people depending upon the foods which she consumed.'

She was admitted to hospital, tested for food sensitiveness, and offered immunotherapy, in addition to regular supplements of evening primrose oil. Her skin will in future be treated with applications of cold pressed natural vegetable oils, rather than petro-chemical-based ointments.

10 July
'Ruth has improved a good deal. Her hair analysis shows that she has low levels of manganese, selenium, copper and chromium.

She will supplement with sodium selenite, manganese sulphate and chromium.

'Her pruritis and eczema still flare up, but her mother says her deafness is improving.

'There have been some setbacks with Ruth's treatment. Although her skin improved, as did her deafness, she appeared to be somewhat overactive. Her mother has therefore stopped it. Instead, her diet is once more restricted and Ruth's eczema has returned, though her behaviour is much less bizarre.

16 November

'This young lady has improved on her immunotherapy, increased in dosage. Her skin and her behaviour are much better, and she rarely complains of abdominal pain now.

'A number of foods still provoke symptoms, in particular pork, cabbage, beans and rice, but she is able to take without problems many foods which previously used to trouble her.'

At her worst, Ruth was intolerant of a wide range of things which she could manage when she was well again.

Alex's problem was similar, but the precipitating factor was not so easy to spot until a holiday abroad a year later threw it into relief. He only reacted to chemicals when overloaded with urban stresses of various kinds.

Alex

3 August

'Alex had middle ear infections from the age of 4 to 6 and has had a good deal of treatment with antibiotics for recurrent tonsillitis. He had his tonsils removed at 7½ and then tonsillar remnants and adenoids removed nine months later. Since then he has had further ear infections and his mother has lately been suspecting that he might have milk allergies. She has taken him off milk and milk products. He says that he himself has not perceived any difference off milk as yet.

'There are many problems that foods induce in him; strawberries provoke mouth ulcers, as does pineapple. This is not related to the acidic quality of the food as oranges are perfectly all right. He gets headaches whenever he has fish, and rarely eats this at all. He also has headaches when he eats beefburgers which are bound with cereals, though he is able to take pure beefburgers without problems. In general when cereals are added to burgers a flavour enhancer (monosodium glutamate) is included, and this may be causing his headaches.

'He also has symptoms of dust and pollen at times with hay fever, conjunctivitis and sneezing. His throat is continually full of catarrh.'

He subsequently proved intolerant of a long list of foods, chemicals and inhalants, to which an immunotherapy vaccine was prepared:

potato	pea	mixed beans
tomato	lamb	tea
formalin	beef	cheese
maize	house dust	wheat
ethanol	chicken	milk
pork	cane sugar	beet sugar
fructose	egg	yeast
peach	plum	lemon
carrot	corn syrup	brussels sprouts
soya	cod	shellfish
mixed moulds	house-dust mite	herring
terpene		

'Alex is very much improved with no further attacks of otitis media, but still complains of occasional noises in his ears. His nasal catarrh is also greatly helped by the anti-allergy treatment he is receiving.'

Five months later . . .

1 April
'Alex has been very much better for a long period. However, lately his headaches have been returning after he eats, and his catarrh has been increasing.' His vaccine was reassessed.

Alex went to Greece on holiday in August. There his symptoms completely cleared. On returning to London some of his symptoms have relapsed, indicating that urban pollution overloads him.

Asthma
A great many people do respond to intolerance problems of all kinds with symptoms in their chests, but medication is often very successful at suppressing them.

The spongy lung tissue itself is neither sensitive to pain nor structured for self-defence; it just gets on with a highly specialized task. But it is very vulnerable, and depends on defences higher up in the system of branching passage ways of ever-decreasing size which eventually lead to it from the voice-box. We could see from a plaster cast of the spaces in the tubes and tissue of the lungs that they closely resemble a miniature tree, suspended upside down with its trunk at the voice-box and the foliage spread around the ribs.

The trunk and main branches of this tree are held rigidly open by rings of cartilage, so any challenge that gets that far proceeds unopposed. The lining can become inflamed though, thickening up and producing quantities of mucus. This uncomfortable condition makes coughing troublesome, and usually develops during chronic exposure to irritant inhalants. But its effect on the speed of air passage in and out of the lungs is very slight.

Only when the smaller branches are reached is resistance possible. There the cartilages give way to rings of muscle, which can and do tighten if irritated. The little air that is allowed through wheezes as it goes, an effect which for mechanical reasons is more pronounced when breathing out. This response rarely develops during infancy, though a wheezing type of bronchitis begins to be seen by a year of age. If this heralds asthma it is usually obvious by the age of two or three, though older children and even adults may develop it for the first time. Happily, it will often cease to be a serious problem during childhood or adolescence, and may vanish completely.

Excellent medicines are now available to suppress the muscular spasm and prevent the inflammatory swelling and bronchial mucus, and most people nowadays rely on them. Nothing like as many of these children are brought for immunological treatment as a result. But drug therapy does sometimes prove inadequate, and we get revealing little glimpses of the problems that can trigger off asthma.

Geoffrey

'Geoffrey is a little boy of 2 whose family live in the country amongst farm land. He had been sent to the hospital by his doctor for troublesome catarrh and coughing at night, and occasional bouts of wheezing, which had come on since they moved there. Tests were done, which revealed nothing strikingly wrong with him, so various medicines were tried. Sometimes one seemed to work, then the symptoms would return in spite of it.

'Mother became convinced after a year of trial and error that medicines were really not much use, and began to feel the same about the doctors. They in turn began to wonder if she was not a little hysterical. The family was thinking of moving house, to get away from the scene of it all.

'Then a friend showed her an article in a woman's magazine about food allergy, and she brought it for one of the doctors to see. He encouraged her to try it, and recommended a weekly programme of exclusion trials to check the most probable items.' [See chapter twelve].

'Six weeks later mother almost danced into the clinic, radiant with success. During the week she had kept Geoffrey off everything to do with the cow, he had been a different child. Once back on it, he had rapidly regressed.

'Nothing has been seen of them for six months since then, and it begins to look as if his problem is solved.'

It seems he had tolerated food from the cow well enough at first, and had been triggered into asthmatic and bronchial symptoms by regular and unavoidable exposure to agricultural sprays. The family could not deal with that except by moving; but withdrawal of bovine food lowered his total irritant burden back into a range he could cope with. He can now tolerate the irritants he cannot avoid, by abstaining from those he can.

If Duncan's problem had taken so long as Geoffrey's to sort out, he would probably have died. He was born fifteen years after his sister, following a series of miscarriages, caused by mechanical weakness rather than mother's ill health.

Duncan

'He was seven months old when he first had wheezy bronchitis. Bouts of croup and wheezing followed through his second year, and he had his first hospital admission for severe asthma three weeks after his third birthday. After another admission at death's door just a month later, the family and their doctor had a conference. If after only two years Duncan already needed full doses of all the medicine available to keep him out of hospital, what kind of a future had he? They decided to try something radically different, a diet as free as possible of chemical irritants.

'The whole family ate only raw fresh food for the next three weeks, to keep Duncan company. After a fortnight everybody had lost some weight and felt much better. Big sister's skin blemishes had disappeared. And Duncan had not wheezed once since coming home from hospital.

'Duncan is now nearly five. He has had ups and downs, and still has his medicines standing by in case of trouble. His parents are content with a compromise. Rather than keep him on a rigid diet which makes visits and tea parties difficult, they are using medicine and supplements of fatty acids and vitamins to enable him to deal normally with life. He has not needed steroids for three years, which is an important gain. His is fully active and growing normally. And if ever he is overwhelmed by junk-food indulgences on outings with relatives, his parents know they can fall back on a fundamental diet for a week or two to clear things up.'

It appears then that serious difficulties in dealing with the world and its contents can be expressed from an early age as intolerant responses on any level in the respiratory tree. Irritants do not have to be inhaled to produce this effect; an irritable tissue is just as easily provoked from within as from without. While the defences are down the world can seem your enemy and everything within it. Yet if you have a chance to get everything back in balance again, these enmities seem to dissolve away. It is scarcely any wonder that medical allergists find intolerance so hard to believe in: when overloaded it seems too theatrical to credit, and when back under control it cannot be demonstrated at all.

CHAPTER NINE
Gut Reaction

In that beautifully straightforward life before birth, food was as meaningless to you as air. Everything you needed was available, plumbed in directly, part of yourself already. Pain or panic sometimes resulted from hostile substances mother exposed you to, but you had no means of relating the two.

Even after birth the discovery of separate food is not so immediate as breathing has to be. If mother feeds you herself, your nourishment retains the character it always had but comes to you now in meals, through a mouth and stomach you had only used for swallowing fluid before. So you are free to accept that food with confidence, and learn one by one the lessons of digesting and assimilating it. There should be nothing here to arouse the defensive reactions you are simultaneously learning in your nose. But they are preparing you for the time when you need to seek food further afield.

Your first tastes from the family table will be strangely different, just like those your nose encountered at the moment of birth. Six months' experience have prepared you well; you can manage now to accept in your stomach substances which would have provoked you to reject them, just a couple of months before. Now you can get down to digesting them, and turning them to advantage as nutrient resources.

That seems to be roughly what nature intended, but unfortunately it has not always worked out like that. If there were moments of pain and panic before you were born, the irritants which caused these are now conveyed to you in mother's milk. Their first impact is now focused on the intestine, still unprepared to cope with irritants of any kind. Lessons in digestion come therefore to be confused with self-defence, and perfectly good food is rejected along with the irritants mixed in it.

Vomiting, colic and diarrhoea are perfectly understandable results.

If mother avoids the irritants herself, the problem can easily go away. But she has to know where to look. She is just as likely, even today, to be told that her milk is too weak or too strong, or defective in some other way. So she may give up feeding you herself, and try one or other of the baby feeds formulated from cow's milk.

That step leads directly from the frying pan into the fire. Whatever may have upset things before, at least the basic food was right. Now it all comes from another animal, intended for her calf. If the calf had instead been made to depend on your mother's milk, it would not have survived. We have our highly adaptable natures to thank that we can get by at all with artificial feeding methods.

We met Michael in chapter one. The early stages of what was to become a perennial nightmare are described in a letter of referral from his doctor to one of his many specialists. Perhaps even in hindsight it would have been difficult to manage any better. But the seat of the chaos is clear enough.

Michael

'Michael started with recurrent episodes of loose stools at the age of seven months, having been born by normal delivery with a birth weight of 4.7 kilos and previously being breast-fed with complemental cow's milk formula. Many changes of diet have been tried, including the use of soya milk, gluten-free diet and the avoidance of various foodstuffs; but none of these appear to affect the frequent loose stools that he was having. Despite all this, however, his growth rate has always been excellent. We had him into hospital in March of this year for investigations and he was slightly positive on skin testing to nuts, but essentially negative to all other allergies at that time and basic investigations did not reveal any other abnormalities. Yet he has continued to have frequent loose stools when challenged with particular foodstuffs.

'I think that basically he does have a gastro-intestinal allergy and there is a family history of this on his mother's side. He has also had mild problems with urticaria after eating certain foodstuffs and possibly some seborrhoea of the scalp.'

It is very easy for the comparatively simple problems of fetal life and infancy to become compounded into major preoccupations, like those Michael went on to contend with. Whenever food and irritants associate together in the same meal, the lesson is more about rejection than acceptance. The more alternatives we try as means of escape from the obvious feeding difficulty, the more trouble we encounter.

Most healthy babies cannot cope well with widely diverse foods until well on towards the end of their first year. Even that supposes that their previous experience has made them well nourished and well constituted, and able to devote a large slice of their curiosity to exploring food. Children with histories like Michael's are forced to take on major challenges prematurely, blindfold and with one arm tied behind their backs.

So from their first complemental feed, children like this build up a repertoire of foods and chemicals they will tend to reject under stress for the rest of their lives. Through anxiety to ensure that they survive at all, they receive far too much solid food early on, to be overwhelmed with quantity as well as quality. Refined sweetened rusks and cereal puddings flood their circulations with glucose, putting their insulin systems under pressure right from the start.

Digestion goes horribly wrong, producing by mistake gases and irritants that do more harm than good. Delicate absorbent linings in the small intestine get involved in inflammatory reactions they were never designed to stand up to. The wholesale wear and tear that can result from this destroys the tissues on which absorption depends. So basic food substances like lactose fail to enter the body, and linger in the gut as food for bacteria or yeasts. Fungi like candida often overgrow because catarrhal illnesses misunderstood as infection have been mistakenly treated with antibiotics. These kill off the bacteria which normally inhabit the intestine and assist digestion, leaving no competitors to keep yeasts at bay. If these prosper too well they can transform into mycelia, enormously extending their potential for mischief.

And the large bowel is supposed to organize all this chaos into faeces! Even in children without inflamed or overactive intestines, there are problems. The deficiency of fibre in much of their food means that residues accumulate slowly; so the organ which collects them is liable to be stagnant. The hard little pellets that form there are difficult to move on, and irritate the skin of the bowel where they lie. The appendix is just nearby, and can easily become congested and inflamed in this situation.

Most appendicitis, tummyache and constipation results from refinement of food. It is one of the commonest group of ailments for which children are taken to doctors. If allowed to settle into a lifetime's habit, they go on to cause a wide range of common, uncomfortable and dangerous disorders from haemorrhoids to gallstone, diverticulitis and bowel cancer.

In active intolerance, the colon may adopt the opposite extreme – it gives up trying, and gets inflamed and irritable itself. Faeces do not form at all. Foul, sour diarrhoea loaded with undigested food is the miserable result, mixed at times with mucus and blood from the inflamed patches of the colon itself.

Elizabeth's long and complicated problems began like this.

Elizabeth

18 November

'Elizabeth was born following a normal pregnancy by normal delivery. Her mother had some hypertension in the last few weeks of her pregnancy. Elizabeth's birth weight was 6 lb 15 oz. Her bowel worked at birth and her lungs were checked to see whether there were any problems, but she was deemed to be well. She was breast-fed for four months having only a single bottle of formula during that time.

'At about two months of age she had three days when she vomited everything. Her stools changed, consistency became more mucoid and she appeared after that to have abdominal pain after feeds and tended to vomit a great deal. She continued to put on weight and looked healthy enough but was given the diagnosis of infant colic. At four months, however, she started passing blood. She stopped putting on weight and was referred to hospital where she was said to have a behavioural problem and her mother was advised to feed her more frequently. However, she appeared to be in more pain and to pass more blood.

'It was then suggested that she could have lactose intolerance and this was confirmed as the stools were loaded with sugar. She was transferred to soya milk and appeared to thrive for a while. Then she developed gastroenteritis and her stools changed from yellow to a pale mustard colour and contained mucus. Baby rice was introduced and she became extremely ill; subsequently she was admitted to hospital just before Christmas and lactose intolerance was confirmed. Thereafter she was seen by a doctor and admitted to another hospital for two weeks. She was put on a solid food diet, including potato, kosher margarine, puréed lamb and puréed apple. She improved until six weeks after that was instituted when she fell ill again.

'She was head banging, appeared to be unable to concentrate and was very disturbed and tearful with abdominal pain. A very basic diet of cabbage, banana and gluten-free rusks was attempted but her condition was again poor. She therefore returned to chicken, cauliflower and apricot and appeared to be all right on these. Each time white bread was reintroduced to the diet, or any cornflakes or grain products, she appeared to deteriorate. She is very hyperactive but appears otherwise to be well.'

She was admitted to hospital on 19 August when she was put on a diet which excluded her commonly eaten foods for the first four days; for the subsequent four days she was challenged sequentially with her commonly eaten foods, one at a time and her reactions were recorded as follows:

Cornflakes	– became very miserable after finishing a bowl of cornflakes. Was happily playing prior to this meal.
Lamb	– became pale after the meal and was lethargic and unable to walk properly.
Potatoes, Carrots	– reactions not conclusive.
Bananas	– became extremely agitated after finishing a banana.
Pears	– became miserable and started crying, wanting to be picked up all the time and pointing to her abdomen; drawing up her legs as if she had a colicky pain.
Chicken	– did not want any more after a spoonful.
Apples	– became extremely aggressive and started throwing things about. Started crying, lashing out at her mother and banging her head against the cot.
Pork	– felt very sleepy after the meal.
Peas	– no apparent reaction.
Plaice	– felt miserable after this meal.
Parsnip	– unable to walk steadily.

These items were puréed in the blender before being served.

'Using the Miller method she was tested to candida and yeast, pork, rice, peas, carrots, maize, apples, chicken, potatoes, cauliflower, fructose, eggs, cane sugar and wheat. She was then started on a cocktail of these in a desensitizing vaccine. The principles of a diversified rotary diet were explained to her mother who was agreeable to following this diet.

'When tested, she definitely had abnormal reactions. She was tired, unable to concentrate on occasions, sometimes wobbly, her pulse rate varied and she became very miserable at times. On occasions she was hyperactive and at times appeared almost manic.

'Elizabeth definitely seems to have food intolerance. However, her mother did not maintain the treatment once Elizabeth came out of hospital. She felt that sometimes the vaccine which was provided unsettled her.'

Unfortunately, her problems are by no means over yet, but have taxed the belief of her parents and the doctors involved. Her specialist's response to an unstated challenge conveys some of the pressure this kind of situation engenders.

22 January
'A case conference was held about this girl. My clinical observations and those of my team of colleagues and staff are that Elizabeth does react to foods shown both by elimination diets followed by challenges and tests using the Miller technique. These observations do corroborate what her parents have observed and indeed others also.

'The basis of the diagnosis of food sensitivity is clinical. Laboratory tests are not as refined yet as to be able to corroborate these. Indeed there is an immunological network of overlapping factors which tend to compensate for each other, so it is extremely difficult to pick up 'in vitro' the responses that occur. I would therefore hold to my opinion, which has not been lightly made, that Elizabeth does react to foods and in addition, this may be a threshold effect with a number of items being cumulatively responsible for provoking symptoms on some occasions. It is very well known in hay fever, for example, when many people may not develop symptoms of hay fever unless they are also exposed to some other agent, and in addition, the symptoms of the rhinitis continue after the end of the pollen season in many people for some weeks. Having been triggered by the pollen, they continue to respond to other agents in the environment which would not normally provoke the rhinitis.

'It seems to me that the problems are threefold:
1   that Elizabeth is managed by dietary manipulation;
2   that her mother has been very tired and worn by the care she has had to devote to Elizabeth;
3   there is difficulty in coordinating all the care that is required especially as private medical help has been sought.

'With regard to the first, I think it would be quite possible to ensure that Elizabeth has an adequate nutrition, and indeed we have two nutritionists who can assist in this, one a paediatric specialist and the other with a vast amount of catering experience as well as her Master's Degree in Nutrition.

'Second, I feel quite sure that the support you have been offering her mother will help her. She needs practical support such as that which you have been devoting to her.

'Third, I see no difficulty in working with colleages in the Department of Health and Social Security and the National Health Service.'

Billy is more fortunate. He is old enough to realize what is going on, and responds to the aversions his senses signal to him. But look at the list, as it stood when he was first tested! With treatment he stands to regain his tolerance of most of these.

Billy, age 10

19 December
'All Billy's family have an unfortunate health record. His father has an over-active thyroid and his mother's is under-active.

'Billy himself was breast-fed and then supplements were included. The breast feeding was stopped at six months. He went through normal developmental milestones, and was using 250 words at the age of 2. By 6 he began to have symptoms. He would have a cold or virus infection and would develop aching of the legs and arms, with headache and fatigue. He used to complain of severe abdominal pain and stopped eating his lunches at school. His mother therefore gave him a packed lunch, which was unusual but permitted in his case.

'His younger brother was taken to an allergy clinic by relatives because he began to vomit after all meals, and Billy was tested at about the same time. The test used was a muscle power test when challenged with the food under the tongue. A diagnosis of allergy to cow's milk, egg and cheese was made in Billy's case; also chocolate, citrus fruit, additives and flavouring. These foods were eliminated from his diet with considerable improvement in his condition in that his writing became better, his concentration improved at school and for the whole of the following year he felt a good deal better. However, this year in February his symptoms started again and he has had to be off school a great deal with recurrent respiratory tract infections, possibly following whooping cough. He has been seen by a paediatrician who has regarded him possibly having had a virus infection and a post-viral syndrome and recommended he should have aspirin. Homeopathists have recommended a number of powders; with the last set of four powders he has been offered there has been some improvement. However, he has not been able to do games or PE because of pains in his legs and arms and he is really unwell in general.

'He is very conscious of smells – strong, scented chemical products. He dislikes petrol fumes, and won't use glues if he can avoid them.

'Billy recently saw a paediatrician who found nothing significant apart from a few enlarged glands in the left side of his neck. He is close to the 50th percentile for height and weight. His father was recently made redundant so there may be psychological factors.

'It was arranged for him to have a rare food diet to identify his problems by re-introducing common foods one at a time.'

4 February

Billy was admitted to hospital and put on a diet of rare foods, then challenged with a number of common foods and substances. He proved to be sensitive to a long list:

carrot	chlorine	house-dust mite
peas	gas	candida and a mould mixture
banana	MSG (in food)	feathers
tea	terpene	cat
egg	ethanol	dog
milk	phenol	pineapple
chocolate	celery	wheat
sugar cane	soya	fructose
melon	goat's milk	lettuce
maize	beet	chicken
corn syrup	rice	apple
onion	sprouts	potato
tomato	cheese	

Dennis turned out to have a simpler problem with a happy ending, though it took a year to track it down. Meanwhile he had to put up with dietary restrictions and still had his diarrhoea.

Dennis

3 January

'Dennis possibly has food intolerance. He has a number of restrictions in his diet, in particular milk and milk products, possibly apples and oranges and eggs. He has loss of bladder control and enuresis when he takes oranges; coloured sweets provoke diarrhoea; and if he has milk he loses his appetite.'

He received immunotherapy and dietary supplements.
A year later . . .

19 January

'Dennis had a return of his diarrhoea this autumn. When he had bottled water instead of tap water or water from plastic bottles, it cleared completely.'

Absorption from his intestines proved to be a bit deranged by chronic exposure to chlorine, and thirteen months later . . .

12 February

'Dennis' mineral analysis shows that he has very low chromium, zinc and manganese.'

He took supplements again, and has regained his health. As we anticipated, nutritional balance goes rapidly astray in situations of this kind, and frequently calls for supplementation before matters will come right.

Diarrhoea is never any joke. Not only is it unpleasant to suffer and to deal with, but the inflammation that gives rise to it can have trying consequences.

In Owen's case, the colitis which was his original problem never did get sorted out. He wanted help with the migraine he gets from the drug he takes as treatment for his bowel. He is in Catch 22.

Owen, aged 9

'This boy attended a neurological hospital with migraine so prostrating that he spent several days a week at home. He had ulcerative colitis and bled from his bowels three or four times a day, so was on a drug to control this. His sensitivity to the drug had resulted in headaches and he had developed migraine.

'This type of chemical sensitivity can be a trigger to food sensitivities. In this child's case school problems ensued.'

For example . . .

25 May
'Owen has had a prolonged illness of colitis and migraine. He has been receiving treatment for migraine for some years, including desensitization to a number of foods. He has improved with this and his headaches are reduced, but he does feel ill at times and is subject to severe headaches when he encounters allergens for which he is not protected. He has treatment for a range of foods and inhalant allergens and in general has to have a restricted diet omitting colourings and additives in particular.

'His medical condition will entail his having some absences from school.'

His treatment consisted of diet, mineral and vitamin supplements and avoidance of chemicals as far as possible.

28 May
'Owen has low zinc and high copper levels shown on his hospital tests. He is also extremely short of selenium, and has started taking a daily supplement of sodium selenite and zinc sulphate.'

One year later . . .

'Travelling with exposure to chemicals in traffic used to result in headaches and prostration but his condition is now steadily improving, so that he can travel with less ill effect.'

The predicament of young people like Owen, with serious chronic inflammatory bowel disease before they have left school, is really very worrying. We do not yet know what faces them in life, but can be sure it will be hard. They certainly face greater risks of bowel cancer, false passages (fistulae) and surgery. Their nourishment is at a permanent disadvantage, so that getting well again is repeatedly frustrated by insufficiencies and imbalances brought on by relapses of their disease. They are prey to anaemia, exhaustion and involuntary starvation. Yet they are members of the most affluent society yet!

Arthur's tale is an extreme one, told to press this serious situation home. The record available spans his childhood and adolescence, and sets him out upon adult life already an old man. His stubborn indomitable courage deserves the homage of all the rest of us, and a better deal from life than has yet been his lot. Problems began in his intestine, and the most obvious scars are there. But no part of his body or mind has escaped the ramifications.

Arthur, aged 20

2 August

'He apparently had difficulty with feeding in infancy and was admitted to a London children's hospital because he had projectile vomiting. However, this gradually subsided and he was able to feed, but he developed frequent episodes of catarrh and had to have many penicillin courses for his throat infection. Adenoids and tonsils were removed when he was five. He had infantile eczema and this cleared gradually, to be supplanted by asthma which was exercise-induced. He was found to be sensitive to wool, feathers and house-dust mite; also dogs. He gradually recovered from this and with elimination of wool and feathers from contact with him or near proximity to him his asthma improved. He was able to go jogging and went on long walking holidays.

'He went to a polytechnic in 1980 and noticed that he was extremely tired – abnormally so, he thought – for the whole of the year until the following autumn. During the summer he felt a little bit better in the open air on a kibbutz in Israel. However, his job there entailed spraying the orchards; the grapes were sprayed with sulphur and other fruit with other pesticides. When he re-started work in the autumn of 1981 he again noticed how extremely tired he was. He had moved into lodgings with student friends, and he used to do the evening cooking, using a gas cooker.

'He had been a huge bread eater until the autumn of 1981 and used to drink a great deal of milk. Now he has an aversion to milk and bread.

'When he was first attending college he was unwell and developed ulcerative colitis which progressed to a very severe debility and within nine months he required a total colectomy (complete removal of the large intestine). During this time he also had disease of the nerves in his arms and legs and was under observation in hospital.

'He still has the neurological disease, so that he cannot feel parts of his limbs and he cannot walk without support. He is emaciated. He has joint enlargement and severe pain in his limbs. His bowels are opening 10 to 30 times a day and he passes blood. He is virtually bed-bound.

'The indications that he has some food allergy are the projectile vomiting in infancy, recurrent catarrh, asthma and eczema. Colitis often responds to dietary measures, and his year of "total exhaustion" preceding the colitis, could have indicated food sensitivity, for he put on no weight, despite eating huge quantities of bread and drinking a large amount of milk.'

He started on fresh food and began to avoid milk and wheat. He used an intolerance-blocking medicine while building up his constitution, so that if he was sensitive to other foods, at least he was partially protected. Blood tests were done.

30 September
'He has a high level of antibodies to wheat, milk and egg. He should have only organically-grown food as far as possible and his family have been pursuing this for him. In addition to this, he is being desensitized to some common foods and he feels somewhat stronger.

Four months later . . .

10 February
'Arthur has been improving on his desensitization treatment and his weight has been increasing. He still occasionally has pain in his arms towards the evening, but in general he feels better. In the past few days, however, he developed a sore throat and an ulcer appeared on the right tonsillar pillar. A throat swab showed a heavy growth of germs.

'In view of his highly sensitive state and his previous reactions to antibiotics, he was given hydrogen peroxide mouth washes and antiseptics. Though at first he developed two more ulcers he gradually improved and now the ulcers are healing. Arthur feels a great sense of accomplishment in being able to deal with such an infection without any serious consequences.'

11 March
'Arthur is improving. He is putting on weight and feels very much better. He has an extraordinary symptom however; from 5 o'clock in the afternoon he gets severe pain in his arms and legs. This reaches a crescendo in three hours and on treatment with codeine it will quite rapidly abate. This could possibly be due to withdrawal effects from codeine, making him crave another dose in the evening.'

On desensitization to codeine all his evening pain cleared.

Ten months later . . .

27 January
'Arthur is improving on his treatment. He is now able to drive, and is undertaking a part-time degree course in computer science.'

22 September

'Arthur rarely has blood in his stools now, although of course, with a total colectomy he still has rather frequent stools. The sensation is returning to his hands and legs, but he still can't walk very far without becoming exhausted. He no longer complains of the excruciating pain in his arms.

'His mental state is worrying however. Because he is unable to be independent he is becoming a recluse. He is a very angry young man.'

The huge problem of returning to a normal life began to occur. Before this, survival was the only instinct. He declined any further operative treatment but had some psychotherapy.

13 January

'His family has been trying to get him to adopt a more outward-going approach. However, he really is incapable of coping with a normal life and has made steady improvement, though slow, since his treatment started. The current acute illness appears to be due to another respiratory tract infection.'

Blood tests actually showed major derangements, including allergic manifestations and anaemia. He was quite ill, and proved to be intolerant of the new dog at home. He got better after banning it from his bedroom.

10 December

'Arthur has been very well for many months now. He is at college doing a course, working seventeen hours a week there and doing a considerable amount of homework as well. In the last two or three weeks however, he has had more of the aching pain in his shoulders and arms and the acne rash on his back. He also finds that he is unwell with the aching discomfort coming on as soon as it rains, and it clears when the rain stops. But his ulcerative colitis has been under control for the last three or four months with no episodes of diarrhoea and no mucus or blood in his stools.'

Thirteen months later . . .

9 January

'His mother says Arthur is 60 per cent better. His bowels are not open more than once every eight hours now and although he is still exhausted and spends a good eight hours in bed every day, and still has neuralgic pains from the peripheral neuritis, in general he is very much improved.

'He has managed a year at college and although he is retaking that year because he only managed to stay for part of the course, he is progressing.'

This boy had multiple sensitivities especially to some drugs – a sulphonamide drug used to treat ulcerative cc'itis, and a codeine drug for pain. He had to avoid gas, perfumes and strong chemicals in his home. By taking all these factors into account and coping with his natural anger about them, he is now leading a relatively normal life.

CHAPTER TEN
No Exit

Our nature provides us with highly discriminating border guards, which normally enable us to reject the grossest forms of poisoning, whether they assail our nostrils or our stomachs. Until recently, what we have admitted to our bodies has usually been restricted to a manageable minimum.

We are provided with plenty of excretory reserve. Not only do our kidneys and liver excrete liquid wastes into urine and bile, but masses of volatile material is exhaled in our breath. Our perspiration affords another route by which oily or watery material can leave us. Anything deposited in the deep, growing parts of our skin, finger-nails and hair will eventually be shed in the surface layers as they fall or wear away. We should have no difficulty keeping dangers at arm's length.

That works well enough for traditional exposures like heavy metals, but comes under challenge with many newer chemicals. We have not had time to learn how to get rid of many of these, and some of them are dangerous enough to poison the tissues through which they must pass to get out. We complicate this even more with the congestive effect that sophisticated modern foods often have. People who subsist largely on rich proteins are jamming their metabolism and circulation with the products of digestion. Swift and adequate elimination is difficult enough for these metabolites, let alone extraneous chemical toxins – with which foods of this kind tend to be the most heavily contaminated.

So some of these irritant chemicals accumulate to produce effects in the body while they wait. Urticaria is a disease of that kind – an acute generalized itch, with blotches of swollen skin which can come on with great rapidity. That was Gregory's problem.

Gregory, aged 2

19 January
'Gregory has had urticaria since birth. His Swedish parents work for an American company and are here for some time. His mother says he has had urticaria with hard white blotches on his arms and red blotches on his face whenever he eats certain foods. He also suffers from cold urticaria.

'The particular foods which provoke his symptoms include colourings and additives and some common foods.'

He was tested, and nutritional blood tests were done.

10 April
'Gregory responded to his desensitization treatment and has been very well. It had to be adjusted a little bit lately, but in general he has had a clearance of his urticaria on immunotherapy.'

8 November
'Gregory's skin has improved completely on his immunotherapy and while the family were away in the summer in clean air. Now that he is back in London the urticarial weals are coming back slightly and he reacts occasionally to apple, grapes, plums, shrimps, soya and tomato.'

An ionizer and air filter were installed in his bedroom.
Ten weeks later . . .

22 January
'Gregory has improved with regard to his urticaria. When he is on his vaccine he remains very well. However, he has difficulty in breathing through his nose. His tonsils are enlarged as are his tonsillar glands and he probably has enlarged adenoids.'

Later that same year . . .

20 September
'Gregory has had no episode of urticaria since April. His treatment was discontinued in July, and he has had no recurrence of any symptoms related to food allergies for five months. There is only one food he cannot have, tomatoes. Otherwise he is able to eat fully.'

If toxins and heavy deposits of acidic metabolites are in the circulation persistently enough, they can contribute to another range of problems. They disturb normal metabolism so much that the body desperately looks around for quiet corners to deposit them, out of the way. Favoured sites are joints, muscles, ligaments and tendons. These tolerate a certain amount, like the land-fill sites we use for municipal rubbish dumps. But eventually large piles of rubbish begin to fester: in the body, they inflame. Rheumatism and arthritis are typical results.

Modern medicine dismisses this kind of reasoning, though it provides no better ideas to substitute for them. It would leave us at a loss to account for Rachel's symptoms. Some of them have a familiar ring, and she found keeping to her diet and treatment extremely trying. The worsening of her condition which resulted each time she lapsed, only serves to confirm that her treatment had a direct connection with the improvements.

Rachel, aged 12

15 September

'Rachel's story started before Christmas when she was ill with a throat infection. After Christmas she was taken to a doctor as she also had diarrhoea and in the meantime was given aspirin. She was diagnosed as having gastric 'flu, but shortly after that her knees became swollen and it was suspected by a doctor that she had an allergy to food. She was prescribed antihistamine and subsequently sent to see another doctor. At that time streptococcal septicaemia was suspected and perhaps rheumatic fever. She was given penicillin. Subsequently she was seen by a further doctor who excluded rheumatoid arthritis and an eye surgeon excluded the possibility of eye involvement.

'Since January, however, Rachel has more or less been unable to go to school adequately. She went back for a few days at a time but still complained of headache, diarrhoea and nausea, and her knees became extremely painful. At Easter her mother suspected that foods might be responsible for her problem and put her on a dairy-free diet. Since then occasionally she has had a strawberry milk shake and has been sick the next day and has had diarrhoea. She tried two intolerance-blocking drugs, but has not really responded to these.

'She was taken to Italy in August on holiday and there for odd days she had headaches and her knees were uncomfortable, but she was able to be active. However, as soon as she returned home to Britain she became wheezy and had to have treatment. She has also had hives on her arms and been much more ill than she was prior to her holiday.

'On examination there was no abnormality to find generally apart from slightly swollen knees and painful tender left wrist and hand.'

She was admitted to hospital for elimination diets and provocation tests by Miller's method, followed by intradermal skin testing.

15 September

'In hospital Rachel went on a rare food diet and then sequentially reintroduced common foods one at a time after a period of five days. Initially she felt extremely unwell and for two or three days she did very little except lie about feeling unwell. Thereafter when we started doing the food challenges she had very definite symptoms. She developed a rash on her arms with pineapple, she had malaise with fish and also with orange, with beef she began to feel sick and complained of swelling in her arms. She had severe headache and subsequently developed abdominal distension so that she had to unfasten her belt. With egg she developed pain in her wrists; apple provoked epigastric pain; with turkey she had a headache and developed a rash on her arms and legs. When she had potatoes she had the most dramatic reaction of all – with severe urticaria on her neck, face, arms and chest. She had nausea, and severe pain in both knees and ankles and the left elbow. She also developed angioedema of the right eye. A similar severe reaction occurred with chicken and peaches. Other foods challenged included pear, which provoked headache and joint pain; grapefruit which produced rhinitis, and there was an urticarial reaction to strawberry and general malaise with wheat.

'Following the food challenges, Rachel was treated for all the foods to which she had reacted, and to some chemicals and inhalants. She will be given treatment based on this for some months.

'She had serum minerals estimated which showed low levels of magnesium and zinc. Manganese was borderline. She is taking supplements of these and a range of vitamins.

'She knows how to prepare her bedroom so that she is less exposed to allergens there, and how to vary her diet so that she is not eating the same food too frequently.'

1 October

'Rachel feels better than she did before she had her neutralization therapy. She is still getting episodes when she has urticarial rashes, but in general these too are subsiding. She is able to go to school without severe joint pains.

'However, there have been quite a number of incidents. She felt unwell in the physics laboratory and reacted with urticaria. In addition, when sitting at a desk which had recently been polished, she found that she couldn't stop sneezing for a while.'

Nine months later . . .

4 July

'Rachel has improved to the point where she does not need any treatment. She no longer has urticaria or joint pain. She does, however, have bouts of hay fever and in the last week has had a fever with a sore throat.'

Seven months later . . .

8 February

'Rachel, having been off her immunotherapy for several months, has started to have symptoms again from last September. During the last few months she has not completed any full weeks at school. She has been complaining of pains in her joints and urticarial rashes and severe fatigue.'

16 April

'Rachel has been on a four-day rotation and on some of her treatment with immunotherapy. She is very much better in general. However, she is still very chemically sensitive. Her friends smoke and a smoky atmosphere affects her. In a chemistry laboratory she has had to leave the room as there are fifty or more chemicals around her. She is not well in the art room. Unfortunately, her teachers are disbelieving, although it is very obvious that she has urticarial rashes. She does, in fact, have pain in her joints and gets effusions into them from time to time. Roughness can be felt in both her knees.'

10 July

'Rachel has had no further urticarial attacks on her immunotherapy for several months now. She still, however, does complain of joint pain from time to time. She is getting more and more unhappy at school. Because she has been off school for so long this is adding to her problems and it is not going to be easy for her to solve her attendance problems, nor to be more compliant about her schooling. However, her parents will perhaps place her in another school for the autumn term.'

25 October

'This is an unfortunate occurrence that keeps happening with Rachel. When she feels well she gives up her treatment and remains reasonably well for about six months, after which she has a relapse of the urticaria and joint symptoms. She has never gone back to the very ill state that she was in originally when she was practically immobile, but she now finds difficulty in gripping pens, and she has had bouts of swelling of her knees. In general she feels less well than she used to.'

Penny had much vaguer constitutional disturbances not just of her joints but of her appetite, energy and general well-being. Whatever it was about corn that upset her, its effects in her circulation were vicious. She is an intelligent and educated young woman now, and can recall the situation vividly.

Penny, aged 22

'For as far back as I can remember I have always felt sick. Sometimes violently so but if not the feeling was still there. It is only recently, since undergoing allergy treatment, that these feelings have become less frequent. Feeling sick had become an everyday event that I hardly even noticed especially as I was never actually vomiting. It is only since I have stopped experiencing this every day that I realize how much energy it took to get through the day trying to ignore it.

'I was always given a well-balanced diet up until the age of 13 when I went to boarding school and I know that I had a craving for sweet and rich foods. The more chocolate and cream on a cake the better. From what people have said to remind me, however, I think my problem really began when I started boarding school and had many different problems. The food was terrible, always the same and full of fat. It was a vicious circle. The more sick I felt, the less I ate and the less I ate the more sick I felt. I would live off crisps, biscuits and chocolate until I went home for weekends when I would stuff myself on anything I could find and vastly over-eat. This went on until I was 18 and continued wherever I went. Sometimes I would go without a proper meal for so long I would be close to fainting and a friend would take me to a restaurant. The first types of food that began to repulse me even at home (my mother is the best cook I know) were cheese and avocados followed by hot creamy sauce.

'I pushed myself to the limit after three very hard years in New York, where I would work 48 hours without sleep and rely on a tin of soup and a packet of biscuits. In the third year when I reached the top of the class and became a respected figure in a very tough school, I began to notice that I was really not well and in the last month leading up to graduation I became very afraid of what was happening to me. One problem was that in New York everyone is so caught up with work and themselves, every time I asked a friend if they thought I looked OK they said yes. By the last week I had to rest every 15 minutes between work. I could not keep my eyes open or focus properly and I needed to be with someone in order to eat. Food just repulsed me and I would cook a meal and throw it away. I could only keep going on chocolate. By now I was vomiting regularly. Not bringing up food, just thick bile like glue, and retching so hard I thought my stomach would split. I now realize that my major allergy, to corn, affects me in this way and the digestion period is 8 to 9 hours so there is no food left.

'My other symptoms were an aching in my knees so I couldn't sleep; a constant need for sleep and the inability to feel awake even after fifteen hours sleep; swollen eyes and itching scalp; a constant and undying need for food – I think I could have killed to get it, it was like a drug; sores in nose that wouldn't heal; irritability and hysteria upon entering a large shop and at various other times; severe blinking spasms; bad headaches; dizzy spells where everything would spin and I would have to get on my knees; blurred vision most of the time.

'My illness had reached its peak when I began to receive the necessary treatment. I had to be very patient and take a term out of school where I was to begin my second degree. However, as soon as I entered the hospital and began to undergo the treatment there, I knew I was on the right track. The test affected me and could make me feel fine one minute and awful the next. My three weeks spent there were the happiest I had had for some time. I was finally in an environment where I was not referred to as neurotic and over-imaginative. My symptoms were finally understood.

'After six months I am feeling 60 per cent better. I still have some very bad days when working with chemicals or coming into contact with smokers but I am back at school. I have had to adapt to a new way of life; no make-up, no alcohol, no very late nights, no long working hours, regular meals. I find it hard to go to restaurants, pubs or parties. However, none of these things is as important as beginning to feel well again.

'Trying to avoid corn is a time-consuming operation as it is in almost all commercial products. This could be an allergy which I will always have, but that doesn't seem important as long as I can control the vomiting and other symptoms.'

Many people at older stations in life have similar problems, though they lack the acute and bizarre behavioural features which forced these youngsters to try anything rather than simply put up with their complaints. It certainly is not 'their age'. And it would take nerve and extreme tactlessness to suggest they 'learn to live with it'!

Getting Rid

The ultimate fate of poisons not deposited in joints is some attempt to detoxify and then excrete them. The liver and kidneys are organs in the front line of this attempt.

I once heard from a horse breeder who lost a favourite animal some weeks after it had grazed in a field that had been sprayed accidentally with a chlorinated hydrocarbon pesticide. At autopsy its heart, liver and kidneys had grossly degenerated to useless, fatty tissue, a finding typical of such cases. They rarely get to court, but this one did. The breeder was required to specify which pesticide was responsible for the accident, a fact the other side were not prepared to reveal. He guessed, and was wrong.

There is human evidence accumulating too, but it is not getting the priority it deserves. A paper was published in 1980 in a learned journal devoted to kidney diseases, which showed that 59 per cent of patients with kidney failure due to inflammatory disease had significant exposure to hydrocarbons similar to insecticides, demonstrable in their blood. For other forms of kidney failure the figure was only 25 per cent. From this and the other evidence presented, an occupational link seemed very likely.

Surgeons in East Anglia have been struck recently by the frequency with which tumours of the kidney are coming to operation. A high proportion of the patients are farmers and their families, which even in East Anglia is not automatic: nowadays no more than a third of the rural population would have close connections with agriculture. The only official record of this is in the proceedings of the House of Lords. Lord Favers encountered the problem through an acquaintance, a farmer's wife who had a kidney cancer successfully removed. He reported his findings at length in November 1984 during his speech to the House of Lords on the second reading of the Food and Environment Protection Bill. Otherwise it has received no official attention we are aware of.

Where kidneys and liver cannot cope and the lungs are inappropriate, the one remaining organ that can possibly excrete poisons is the skin. Whether it does so by perspiration, or by depositing the toxins in skin cells as they are formed, the result is effectively the same. Its structure and function are intoxicated, and the group of conditions arises that we know as eczema or dermatitis.

Why this notion of the cause of skin disease is not more widely held, is a mystery to us. It has been available for long enough, and in terms of its predictions it works well. However, doctors tend to be fascinated with labelling and classifying diseases, because of the way they are trained. If they look further into the causes of disease, they tend to ask the questions they can answer with fashionable modern equipment, such as electron microscopes. Simple biological observation of whole people is very much out of fashion. So they are inclined to overlook or undervalue the obvious, which is most readily revealed by these unfashionable means.

In every one of the cases which follow, excretory overload fits in well with the facts. Consider this snippet from Philip:

Philip, aged 12

'My eczema started with a big red band of eczema round my forehead. This band only shows when I react to a chemical; my face goes very white and the red band gets quite hot.'

This child's symptoms started in April 1985 when he was under stress with examinations. He was also overloaded with Easter foods. In addition he had 'flu. All of these factors led to the onset of severe eczema which became infected. The extreme eczema has been provoked every time he goes into shops, in school chemical laboratories and when exposed to air conditioning.

Breakdown under extreme stress has left this child vulnerable to lesser stresses he previously coped with, so that his excretion is overloaded and spills into his skin.

Angus evidently had a problem disposing of food colourings.

Angus

'Angus's mother has been anxious about his skin. When he was two he had an extensive rash, and after eating strawberries and whipped cream it happened again. When he went to Greece two years ago his rash became worse. It was treated with Calamine lotion and cleared. He has always had sensitive skin on his cheeks and on his buttocks. This has been particularly sensitive when exposed to the sun. When he is sunburnt he gets a rash very easily and unfortunately it does not clear rapidly; it tends to be prolonged throughout the exposure.

'His mother has very sensitive skin also, but other members of the family have no problem. He has very dry, flaky cheeks. The skin on his abdomen is flaky in patches and pigmented, and patches of the skin on his bottom are dry too. His mother has had him on a diet cutting out colourings as far as she is able, and there has been some improvement in his skin.

'Angus has apparently also been somewhat overactive much of the time and his mother reports that he has problems in concentrating.'

A diet free of colourings and additives and dairy food was suggested. In fact, he improved on a diet without food colourings alone.

Jennifer's problem is evidently not so simple, and she required more sustained effort to get it under control. But as her skin improved, her mind cleared too.

Jennifer, aged 3

7 August

'Jennifer was born at full term by a normal delivery and breast-fed for eight weeks. During this time she was well. Then milk and other foods were introduced and the eczema began when she was about three months old. Her parents stopped cow's milk and put her on goat's milk for about a year and then they stopped giving her eggs, but the response to this was not clear because she was in a frog plaster for congenital dislocation of the hips for about ten months. She has been out of the plaster now for the past year but her eczema has been very severe. She has been on medicine for it, and local treatment of cream and white soft paraffin. She has been seen by a specialist who said to keep on with the diet that had already been tried. Jennifer, however, is extensively covered with eczema and scratches so that her bedclothes are stained with blood most nights. Now her parents have bought her complete suits so that her hands are not free, nor are her feet.

'They do know that a number of foods really do provoke her eczema. Bananas make her ill and tomato juice provokes eczema.'

Her parents put her on a diet avoiding colourings, additives, tea, coffee, sugar, grains, and milk. Instead, she had plenty of root vegetables, fruit, meat and fish. Challenge tests were performed and a vaccine prepared.

5 October

'Jennifer's skin has gradually cleared on her vaccines. Her mother says that her mind is also very much better; she can sit and play for two or three hours without requiring attention and in the past two months now, since she has been on her antigen therapy, she has only required to have three applications of cream to her skin. She is also off medicine.'

Jennifer started on evening primrose oil and zinc.

29 January
'Jennifer continues well on her treatment. She no longer has medicine, though occasionally she has an application of cream for her skin.

'She should have a sago-based vitamin C, and alkali salts should she have an acute reaction. In general not only has Jennifer's skin improved, but also her temperament is better. She still occasionally wakes crying, but no longer scratches as she used to. Her parents are pleased with her progress.'

12 June
'Jennifer has had an acute flare-up of her eczema. This happened within an hour of being given a piece of bought walnut cake. In desperation the parents took her to their local accident and emergency department and Jennifer was given medicine, which gradually settled her.

'The eczema is slowly improving but is still present. Her parents have now reduced the medicine. She also had some ice cream at the time of eating the cake, but Jennifer has previously had ice cream with no problems.

'Apparently Jennifer also seems to react when she comes into contact with her grandparents, who are smokers.'

We do not know whether it is circulating irritant that confuses her, or material deposited in her nervous tissues. Nor is it immediately obvious what exactly it is about irritant foods which interferes with skin. Nutritional impairment is one factor no doubt, and agricultural toxin residues another; but these anecdotes strongly suggest an aggravation more specific than that. For the present our scientific knowledge provides no answer.

Eczema is one of the really common complaints among European children today. Hospital clinics are full of them, and they together consume vast quantities of quite potent pharmaceuticals which mask but do not solve their problems. The cases quoted here have many differences, but one similarity: they all found ways to do a lot better than that, which did a great deal more to restore their general health than any medicine ever could.

It must be time to take methods like these seriously, and face the alterations in our ideas about disease that they force upon us. Otherwise we could let ourselves slide into a terminal condition of hopeless neurological and mental irritability. For our brains are more vulnerable than any of the tissues we have considered yet.

CHAPTER ELEVEN
Something on their Minds

Of all the effects that irritant chemicals in our circulation may have on us, it is disturbance of our brains and minds that we have most to fear. That would be nothing short of a living nightmare. Imagine it for a moment.

It descends like a cloud without warning, as if the main fuse has failed. You are separated from the experience of your surroundings. Nothing smells right. Nothing is the right size. When you go to move a limb, the wrong thing happens. When you want to speak, a scream comes out. Your body moves itself around, drunk with violence, a machine beyond control. It crashes into things you know are there but cannot register or judge. All the connections of your mind with your body have been scrambled and make no sense. You cannot control your bladder, though you hate being wet. Passion and panic flood through you in waves, making you yearn and fear without relief of any kind. Nothing satisfies. Then suddenly the cloud can lift, and everything returns to normal.

Experiences like that are becoming commonplace. Most people now know of someone who claims their child is hyper-active. Perhaps some find it a convenient label to explain strange behaviour, just as it was once fashionable to describe every headache as a migraine to indicate how severe it was supposed to be. If it is real, you know it! Ask Jane.

Jane, aged 10

23 July
    'At the age of 5 Jane had migraine roughly every three to four weeks. Her migraine has been preceded by a period when she has become gradually more and more angry and irritable, and she has been almost unmanageable at times with aggression. Her mother has

found this almost unbelievable, when she is frequently an amenable and helpful child. She complains too that she has giddy episodes and can hardly move the top half of her body when she gets ill with a headache. She says that obviously the room isn't spinning round her, nor is she spinning, but she feels so unsteady if she moves the top half of her body that she simply stays still until the episode has passed, which usually takes several hours. Jane also reports that she feels nauseated from time to time, especially when she has some foods, particularly milk.

'She had meningitis as a child and when in a children's hospital she was diagnosed as having lactose intolerance. However, milk was reintroduced into her diet and she seemed to tolerate it satisfactorily as a toddler.

'Her mother sent some of her hair to be analysed for food allergies. This non-medical method involves divination using a pendulum, but the report came back that she was likely to be sensitive to milk and Jane was therefore taken off milk and milk products. She improved dramatically on this regime and confirmed her mother's suspicions that foods might be responsible, not only for her moods but also her migraine. Unfortunately, however, she has since then been taking a homeopathic remedy and most homeopathic remedies are based on lactose. She has been deteriorating lately and does not feel the benefit of her elimination diet and it could be that these tablets are responsible. Alternatively, it could be that she has been developing sensitivities to other items in her diet.'

She was tested and an immunotherapy vaccine prepared.

4 September
'Jane was very much better on holiday with her immunotherapy, but she still gets a little bit cross at times, with some headaches.

'The week before she was out without mother and ate one scone, also lemon sorbet, which seemed creamy. Her mood was extremely aggressive and angry most of the time – very difficult to control.'

She was tested again and treated for a few more items.

12 October
'Mother says Jane is now a different girl on the immunotherapy. She is doing well at school and she has not complained of headache. She is much more cooperative and amenable and mother feels that her behaviour has settled down to a calm normality.'

Four months later . . .

8 January
'Jane has remained well.'

The victims learn to dread the inevitable march of one symptom into the next. Ben knew he was going to be hyperactive an hour before it happened.

Ben

19 December
'Ben was a schoolboy who could not control his bladder or bowel, by night or by day. During an admission to hospital he was put on a rare food diet and subsequently challenged with foods. When initially he was put on the rare foods, he became dry all day, but as soon as challenged he had a number of symptoms. For example, cod provoked headache, a challenge with pomegranates produced headache, abdominal pain, flushed cheeks and a rash in twenty minutes. One hour later he became hyperactive. Beef provoked marked increase in pulse rate and later a rash on his cheeks. Chicken also provoked a rash.

'He had a number of investigations undertaken. These show him to have a low serum zinc level. He should have calcium pantothenate, dolomite, vitamin A and E and vitamin C, as well as the zinc supplement.

'He has been tested to a wide range of antigens and on treatment he is dry and his behaviour is not hyperactive.'

Imagine having two of them at it, from before they can speak! Few twins are breast-fed successfully for very long these days, and this mother obviously tried. But the problems began with the kind of feeding difficulties we have by now seen many times before.

Dick and John

7 December
'Their mother is at her wits' end as they are hyperactive and aggressive. They are identical twins who were born after a normal pregnancy at full term by forceps delivery. Dick's birth weight was 6 lb 6 oz, John's 6 lb 11½ oz. Mother was extremely good during her pregnancy – she did not drink or smoke though prior to that she had done. She had had a miscarriage just before, and she therefore had a

suture put in to retain this pregnancy. She had had a cone biopsy three years before for abnormal cervical cells.

'Following their birth, the twins were breast-fed initially and supplemented with occasional bottles. They were changed to bottle only at ten weeks. Dick tended to vomit when he was given soya milk – his mother tried to keep him off cow's milk, and likewise John was put on soya milk predominantly. Then at three months they were both put on ordinary baby milk.

'They both were early developers; from the age of ten months they were extremely over-active. They began to be aggressive towards each other by biting and scratching each other and they have very frequent tantrums. They are totally unable to be controlled from time to time. When they have been in nursery school they have been so overactive they have wrecked and disturbed other children's activities. They cannot be separated when they are attacking each other without injury to adults who get between them. They have attempted to stab each other and wield implements at each other as well. However, they cannot sleep without each other's presence and their parents have had to remove all objects from their bedroom and lock them into the bedroom at night for fear they will get up and escape from the bedroom and find an implement with which to hurt each other. They have been covered in bite marks and scratch wounds. Dick had scratch marks from fingernails on the backs of his hands and bite marks on his arms. John had some bruises also. If he can't get John when he is in a tantrum, Dick bites himself as a second choice! They are both wet at night and during the day, though they sleep very soundly when they do get off.

'They are well coordinated and have no neurological deficit.'

Neil's mother seemed to be getting along nicely until an accident nearly killed him. The stress overload of this incident made it look for a time as if he had suffered physical brain damage – all the signs were there.

Neil

21 November

'Neil was born following a full-term delivery with a birth weight of 7½ lb. However although he developed normally for the first six months, at this age he was apparently tipped up in his pram and suffocated when he lost consciousness. He had collapsed lungs and was treated with an antibiotic for a year. During that time he was extremely irritable and ill but he gradually recovered. He had normal developmental milestones until the age of 6 when he began to have epileptic fits. He was put on drugs and was in hospital for behavioural problems also.

'He was found to have problems with his adenoids and was seen by a paediatrician in a hospital where his adenoids were removed and an undescended testicle was restored to its normal position. He was again admitted to hospital with his overactive problems. Gradually, his fits were brought under control and he was put on a milk-free diet. In 1982 he reacted favourably to this and his drugs were reduced and tailed off. He has had no fits for two years.

'His mother says, however, that he does continue to have severe headaches, especially when travelling in cars, and that he has psoriasis which is aggravated from time to time. He has patches of psoriasis round his left ear and on his trunk.

'Neil was admitted to hospital for a five-day period of exclusion diets putting him on rare foods only. Thereafter common foods were reintroduced sequentially and his reactions were as documented below:

oranges   – slight blurring of vision

pork   – blurred eyes after 15 minutes, cleared after 5 minutes. Pulse dropped 40 minutes after meal

apples   – blurred eyes immediately; slight headache

eggs   – blurred eyes during meal and afterwards on and off for 1 hour

bread   – headache at 20 minutes, cleared at 40 minutes. Raised pulse immediately before and one hour after meal

chicken   – hot and flushed 20 minutes after meal, cleared 50 minutes after meal.
Raised pulse – 34 points 20 minutes after meal
22 points 40 minutes after meal
16 points 60 minutes after meal'

Tests using the Miller method then formed the basis of an immunotherapy vaccine for him.

If they cope well with weaning to baby foods, some children will then have problems with the wider range of food available to toddlers. The additive content of foods for young children is controlled to some extent, so that the full force of food chemicals is postponed. Christopher's problems began at this stage.

Christopher

22 June

'Christopher is somewhat overactive. He has tantrums and bad temper and gets very hysterical, losing control of himself. His parents thought it was very likely that this was due to foods as he has been much better on a diet avoiding colourings, additives and chocolates.

'He was born after a normal pregnancy by Caesarian section as he had a breech presentation. He was breast-fed until he was eight months old and at five months solids were introduced. His bad behaviour started when he was about a year old. His parents thought it was just two-year-old tantrums but his behaviour problems have continued to be troublesome.

'Apart from the behaviour problems, he also suffers from an awful lot of wind and indigestion and he has continuous colds. He has had croup from time to time and has had to be in a humidifier with steam surrounding him. Antibiotics such as Erythromycin have been satisfactory in controlling his infections, but he seems to have become more overactive and disturbed at this time.

'He has two dogs and a cat at home. He continues to be wet at night; he is dry if he is lifted but if he is left he gets extremely disturbed and unhappy.

'His parents will continue to avoid colourings and additives, do a chemical clear-out of his home, and avoid grains and milk for a short while.'

These children have all been spotted early and given effective help. Others who are not go on with escalating problems, their parents more and more dismayed. Things often come to a head when school begins, and the child seems unable to cope with learning in the normal way. That was the point at which Hazel's parents realized that something was definitely wrong.

Hazel

'The general impression in June 1984 was that Hazel is hyperactive with sensory inattention. Articulation and motor coordination were poor for her age. She would easily throw tantrums when she was unable to understand things around her and also when she was frustrated.

'Her mother initially kept a food symptom diary and from this it became evident that Hazel's behaviour was related to foods she was eating. She was put on a special diet excluding certain items and within a very short time her condition improved. She was noted to be calmer and her attention span was slightly better.

'She was tested for food sensitivities and appropriate immunotherapy was given; also some dietary advice, to be followed with other aspects of management.

'Hazel over the last eighteen months can definitely see an improvement in her condition. In general she is calmer and her speech is clearer; coordination is better, especially if she does not have to rush through everything. Her concentration has also improved. She is not doing very well in school and special facilities to help her in her lessons would be worthwhile. Her parents are caring and attentive and have always done their best to help her in her development. Her response to this regime of treatment has strengthened the hope that further improvement can be expected.'

She is improving at present, but will face new aspects of the problem as she grows up through school and wants to be able to behave like her friends. That is Janice's problem now. When she is well, it is hard for her to accept that she cannot eat as they do. And she is too big now to be controlled all the time. If she chooses to misbehave, there is no one around to stop her.

Janice, aged 14

6 February

'Janice as an infant used to have a lot of abdominal discomfort and by the age of 3 or 4 she was having very frequent tantrums and would not respond to guidance or chastisement. When she was 5 she began to have urticaria and her parents were advised to look into what she was eating, which might have caused it. By the age of 6 or 7 she was very overactive. At this point, her parents were advised about Dr Mackarness' Stone Age Diet and when Jennifer was put on this, there was an amazing change in her behaviour. She became amenable and cooperative. She could reintroduce various foods with impunity, and provided preservatives and colourings were kept away from her she remained well. If, however, she takes a colouring agent she shouts and screams and throws herself about. Her mother says it is almost as though there is a wall between her and everybody else, and she is totally irresponsible to social requirements.

'She has sometimes had problems with cheese, though her mother is not sure whether this was the cheese itself or dye in it.

'She is at boarding school now and can cope with a diet there reasonably well. However, she tells me she has frequent headaches, two or three times a week, and that these occur when she takes something which disagrees with her. She also complains of double vision at times, and has abdominal discomfort.'

Her tests showed intolerance of: wheat, yeast, milk, egg, oats, maize, rye, chicken, cheese, barley, rice, molasses, carrot, cucumber, tea, beet, cane, honey, goat's milk, grape, mixed beans, lettuce, tomato, potato, beetroot, soya, sardine, candida, nystatin, apple, orange, banana, cod, lamb, cabbage, nuts, beer, fructose. Vaccines were prepared.

25 April

'Janice improved. She is able to eat a full range of foods with her desensitization treatment, and no longer gets headaches. Her eyelids still become a little swollen at times, but she has not had any temper tantrums or other emotional disturbances.

'Her blood tests show that she is deficient in zinc and magnesium, of which she will take supplements.'

8 June

'Janice has been very well on her immunotherapy. She is able to eat all the foods which previously had troubled her, though she is still best on fresh foods rather than foods which have additives in them.'

18 August

'She has been reasonably well with desensitization treatment. However, she still has reactions to a few foods and we discovered that these are the ones which contain butylated hydroxytoluene, butylated hydroxyanisole, and benzoates. She has been very well apart from this.'

15 months later.

12 December

'Janice has been fairly well until recently when she started eating junk foods with all the additives and preservatives. For example, after eating a chocolate and toffee bar she started screaming around the house and was generally very irritable, and her mood has been affected by this. Certain chemicals have also been making her unwell, for example house sprays, felt-tip pens, tobacco smoke and perfumes.'

We have already met Michael several times. His case was notable for showing how the drug Disodium Cromoglycate (see page 14) can reduce the effect of food allergies. When the time came for him to go to school his parents decided to educate him privately, seeing very little prospect that anything else would work. They soon found it would be difficult in any case. Mother wrote a letter to their doctor about his first months at school, in which she poured out paragraphs about the difficulties they were having. Here are some excerpts:

14 February

'Michael started full-time schooling at a private primary school on January 13. There are fourteen children in his class. Initially he had school lunches, as do most of the children in the school, and all in his class. He started school on a Wednesday: for the three days of that week he was fine. The following Monday he started to have very heavy stomach pain, crying a lot at home. The weekend of January 23/24 Michael kept baring his stomach – to cool it – he has continued to do this at infrequent intervals. The nights are the worst. He would wake between three and six times a night, usually sobbing with pain, and other inexplicable discomforts.

'For instance, he complained of his head feeling funny – when I asked him "What sort of funny?" he cried, and said it was "getting bigger"; he also has been trying to "let the air out of his ears". His eyelids are permanently swollen, and under his eyes he has a bruised look with the lines appearing quite prominent. Since starting school he has been on double antibiotics, except for three days between one infection and another. He has constant catarrh which smells. The week of January 18–22 he was on school lunches – thereafter I sent him with packed lunches – the pain has reduced, but it is still there.

'Since Monday January 25 he has been on Cromoglycate by inhaler as well as tablets. He constantly coughs to try to breathe more easily. At the moment he has a slight cough which is due to having had a cold, as well as a dry cough to allow him to clear his tubes. He has also started making strange movements in his throat "to let the air in". The other day when I collected him from school he was most distressed because he couldn't cough, and seemed to have difficulty breathing, although I didn't notice any wheezing. He really is having difficulty coping; every few days his body really can't seem to cope, and he runs a temperature.

'Because he is so depressed and not at all his usual ebullient self I keep him home. On 27 January he was at home ill, and had a temperature and was generally under the weather (when he saw her). He starts to recoup over the weekend and then from Monday begins the downward path again.

'I can't continue to take him out of school. He is perfectly capable – mentally – of staying all day, indeed could do with more stretching, and I have always tried not to make exceptions for Michael. Nights are still very much a problem, even over the weekend, although not so bad as during the week. His sleep pattern has reverted to pre-Cromoglycate form. I also can recognize the several frenetic days and nights followed by a very subdued day and exhausted sleep at night, when he may sleep through. He is also having a lot of nightmares.

When he comes home from school his eyes are heavy and drugged looking.

'We've come so far with Michael while on Cromoglycate – but I can see all this slowly but surely slipping away. Each day at school is a struggle for his system, and quite definitely he is feeling less secure. He gets distressed by very small things. Also he is much less even-tempered. His speech is also reverting, in times of stress, to basic words. His assessment showed that his vocabulary was not as good as his performance in other areas, yet his nursery school has confirmed that in their experience he was a most articulate child.

'So where to from here? I can't stand by and allow the work of the last two and a half years to crumble slowly. There has to be someone who has an answer somewhere.'

If the experiences of older children are anything to go by, things go on getting more and more difficult unless effective means are found to control them. That can be an almost impossible task, once the natural order and self-possession of the child's affairs have been seriously disturbed. Progress through school involves exposure to formalin, laboratory chemicals and art and craft materials; and increasing degrees of self-discipline and personal organization are expected of the pupils.

With Bob, an able teenager, we encounter the full battery of problems in their extremes. In addition to wide-ranging food and chemical intolerances, we find him unable to tolerate electrical fields of certain frequencies as well. There has for some years been a suspicion that some people living near overhead power cables or transformer stations are affected by them.

Bob, aged 13

1 June

'Bob was born after a normal pregnancy by a rather swift delivery, and breast-fed until he was eight months old. There were no supplements of cow's milk given. However, with the introduction of solids at about six months he stopped sleeping well – he didn't sleep at night but cat-napped during the day. This continued until 13 months of age. As a toddler he began to be aggressive towards other children, hitting them frequently. He was a head-banger when he was cross. He went to a Montessori nursery school where, whenever he got angry, his behaviour was bad towards the other children.

'He had recurrent respiratory tract infections when he was a toddler and had pneumonia and bronchitis also. He was treated with anti-biotics almost continuously.

'Throughout this period he had frequent tantrums at home and he would get extremely angry. He had a high IQ nevertheless; this was assessed on account of his behavioural disturbances, and it was said to be 149. Academically, he is quite able but he continues to have tantrums at home.

'He had a road traffic accident and was admitted to hospital briefly. At this time he was seen by an ophthalmologist and checked at a hospital for possible epilepsy. After referral to a children's hospital this was discounted. It was thought, however, that the vague loss of vision and the headaches were possibly due to migraine.

'It is known that some foods will provoke him to irritable be-haviour. Coffee provokes not only tantrums but also migraine, and fizzy orange juice will do the same. Brands of aspirin have provoked irritability.'

27 June
'Bob was admitted to hospital on July 1. For the first four days he was put on a special diet avoiding the most commonly eaten foods. On the fourth day of this diet his parents described him "better than they had ever known him"; his temperament was much better and he was able to speak very calmly. On the fifth day in hospital he was challenged sequentially with individual food items. The two foods which produced the most dramatic reactions were milk and maize. When he was challenged to milk he developed headaches and back-ache and felt very tense and aggressive. His speech was unclear and in slow motion. There was also a pulse change of 30 beats per minute. He developed very similar symptoms when he was challenged with maize. Challenges to other foods did not produce such dramatic reactions.

'He was tested to a wide range of inhalants and foods using the Miller technique and immunotherapy was then started based on the results of these tests.'

13 August
'Bob has been slowly improving on his diet and vaccines. He does now know that if he has a food which affects him adversely, he will have 36 hours of misery. He has reacted to corn and still reacts badly to cow's milk as well as goat's milk. As he has lost some weight on his diet he should have a rotation diet which allows him to have the same foods for three days but then to have a rest for four days.'

Bob's problems were then complicated by emergence of the inherited disease Gilbert's Syndrome (see page 130). Apart from giving him bouts of jaundice whenever he came under severe stress, it caused his liver to swell and become tender – stressful in itself.

29 October

'Bob is at last improving. He has a vaccine for a number of common foods, and a rotation diet to minimize his constant exposure to the same foods – that is a three-day-on, four-day-off rotation – and he has now regained the weight that he lost when he was in hospital on a rare food diet.

'His family are thinking of taking him to Cornwall to stay in a "clean" house, where there are no man-made materials used in construction or furnishing. The materials are all hard surfaces such as ceramic floors, hardwood doors, brick and plaster walls. The furnishings are simple and are mainly of cotton fabrics. No scented materials are used in the household for cleaning purposes and the food provided is organically grown; the water filtered free of chlorine.'

Bob improved in Cornwall. On his return to London he was able to discern quite clearly which items were provoking symptoms in him and this, in effect, is an oral elimination and challenge test with built-in results. He knows that he cannot tolerate the chemistry laboratory in school; the physics room does not suit him, as very often there may have been gas burners on in the room. He says he first blundered in the school when he broke one of the Winchester bottles the first day he entered the chemistry room. He seems to be clumsy and uncoordinated when he is in that room, but he knows now that this clouding of his mind and sapping of his abilities is secondary to his exposure to the chemicals. In the physics laboratory he finds he can cope, but the art room seems to provoke symptoms and he thinks that he cannot tolerate the chemicals used in the paints. He cannot cope with the music room easily. He knows that some of the masters at the school wear quite a lot of after-shave and he finds it difficult to tolerate their presence close to him.

He was assessed by a doctor from the Department of Electrical Engineering at Salford University, for his possible sensitivity to electrical equipment and the following comments were made:

'When exposed to a television he gets a headache only if he has been triggered by something else first. When he is in the music room or physics laboratory he gets a headache very quickly within five minutes and his mouth salivates. The music room is a prefabricated room containing a synthetic carpet and the room has also recently been painted. When using the Spectrum tape machine the load-off tapes failed to load for him, although the machine had worked well in the shop. Often when playing games on a computer he makes errors when pressing the various buttons. He has an electric blanket to ease the abdominal discomfort from his Gilbert's Syndrome.'

This documentation about electrical equipment and exposure to electrical fields is important in some allergic patients as there is now evidence that several of the frequencies to which allergic people are exposed are not tolerated well. This seems to be borne out in Bob's case.

'In the music room there are two items which he finds difficult: the new artificial fibre carpet, and the materials of which the room is constructed. The room is prefabricated, and includes a good deal of chipboard which will outgas formaldehyde. This is one of the most potent indoor sensitizers, and indoor pollution from formaldehyde and other materials is considered to be much more serious than outdoor pollution. It is generally one of the main factors in the "sick building syndrome".

'It may be too that a lemon-scented cleaning product which is used in the school is a problem to Bob and it would be preferable to use unscented cleaning products.

'Now that Bob has an understanding of his problem and an insight into what is provoking his symptoms, he will be able to achieve his potential much more readily. He will have setbacks but he is a very intelligent boy and with support should be able to improve and recover.

'While he and his mother were in Cornwall she noticed a great deal of difference in her own condition and feels that she may well have some food sensitivities herself. She feels that as Bob will be undertaking a rigorous regime for improvement in his health to try to obtain optimum health, she would like to undertake a similar regime for herself.'

Cases like this pose entirely new problems for their victims and for society, and it is tempting to cast around for a new drug or desensitizer which will simply make it all go away. That would miss the point entirely, and leave things to go on getting worse.

The stories in Part Two represent the thin end of a wedge. At present, only the very sensitive and vulnerable are breaking down under the accumulated effect of many pressures, all of them undermining life. Unless those pressures are soon eased, many more of us will lose our health to them. Already people are doing so, but do not realize until motivated to change things for the better.

People have lived for fifty years in dread of the atomic bomb, which is an obvious threat to the future of mankind and the world. To prevent that catastrophe we are individually helpless, or so it seems to most of us. We are left to hope that our national leaders can find the wisdom and statesmanship to refrain from dangerous gambles.

Nature seems to have a much subtler fate prepared for us than that. The species that grew proud enough to take the world under its control and choke it with fumes is being allowed to choke itself. The effects are not lethal, but bring about a kind of debility and incompetence more stupefying than madness. Its victims can be dangerous, homicidally so. Or maybe they simply cease to function at all, in any coherent way.

Some may see in this a self-healing terminal disease for the ecological madness of mankind. They forget that it does not strike the madness at its source – hypersensitive statesmen are, if anything, less to be trusted with bombs. Surely the point is to read this insidious change correctly in the first place, and take it seriously now. It shows no sign at all of going away. At least in this scenario we are not individually helpless. Everyone can rebuild life for themselves, starting here. And as each of us does so, we are doing it for everyone.

PART THREE

THE REMEDIES

CHAPTER TWELVE
Discovering the Flashpoints

By now you may well see one of your children in the descriptions we gave in Part Two, and very probably dread ending up on complicated terms with his food and surroundings, denying him adventures for fear of debilitating symptoms. So it is time for the good news.

There is much you can do to reduce your children's contact with irritants, to eliminate from their bodies most of what they have already collected, and to maintain their vitality despite whatever irritation is left. When you set about these measures their symptoms will very likely lessen, or even vanish completely, in a matter of months. At the same time their personality and mental and bodily functioning are likely to improve to their best ever.

This will be hard for them if they are singled out for separate treatment. You will need to join in with them. And, even if you think you are perfectly well and functional now, you will be surprised to find that you improve as well – not so fast perhaps, and not without an obstacle course early on. The struggle to excrete poisons may upset you temporarily (see chapter ten), but the rejuvenation which follows success is a great joy. And the better preserved you are now, the simpler your task and the better your prospects of complete success.

As your first step consult not a specialist, but your own experience. Most people can apply simple methods of deduction to trace their own family's troubles to their source, once they are aware of the possibilities. Take your time, be methodical and thorough, and keep a careful log of your discoveries.

The Exclusion Trial
Allergy to a drug or to a seasonal pollen is usually obvious. It has a beginning, when the symptoms suddenly interfere with

previous well-being to strike a clear contrast. Something like that could be seen in many of the cases described in Part Two during infancy, at the very beginning of their consumption of cow's milk. But it was more complicated than seasonal allergy even then, because the symptoms could be confused at first with simple unhappiness or routine discomfort. So even at its simplest, recognizing food intolerance is not as easy as you may have expected.

Disentangling from all the other ups and downs of life the effects of a foodstuff or chemical your child consumes every day can be even more difficult still. How are you to know whether his nose is running because of infection, the cold weather or an intolerance problem? Your suspicions are only aroused if a symptom is chronically persistent, far too protracted to be anything else. It may not be even that simple. Your child's problems may vary, or be distributed between several apparently unrelated parts of his body – tummyache, joint pains and nervous irritability, for example.

To make sense of this situation, hospital doctors can resort to a simple trick. They remove the subject as completely as possible from the surroundings and diet he normally experiences to see if he improves. Often that produces no dramatic change in itself, however, because he may still have to struggle with residues of his previous exposure for weeks and months after that exposure ends. But if during that period he were exposed afresh to one of his previous irritants, the doctor would see a new and clear-cut beginning. He would be in the same position as a hay-fever sufferer meeting the first grass pollen of each new season. So, within seconds or hours of the new exposure, much more obvious reactions may occur.

These can be very subtle, such as changes in pulse rate, blood pressure or redness and swelling of the skin in the mouth. Or they may be full-blown bouts of the symptoms for which the subject was admitted. However dramatic or subtle they may be, doctors are learning to accept these reactions as signs that the subject cannot tolerate the substances which produced them. So these reactions can be used to test, one at a time, a whole series of individual features of the subject's usual surroundings and diet, noting what follows in response to each exposure. The observer does not look for benefit after withdrawal of an item, but obvious worsening when it is reintroduced.

Up to now, this kind of investigation is not routinely available in every region of the country; as we explained in chapter one, it is still labouring for full recognition. That should not put you off, however. Though you cannot remove your child totally from your home, you can organize his exposure to many of its features so as to get much the same information together.

Instead of changing everything at once, you can alter things one category at a time, leaving everything else exactly the same. The categories must be chosen carefully, which we will help you with in the next section of this chapter. But once they have been selected and planned out, it will take you just one week to test whether each category gives your child problems. You simply exclude from him every item in the category for the five weekdays from Monday through to Friday, then reinstate the category on Friday evening as if nothing had happened. If you have a bad weekend with him, that category contained items which your child cannot tolerate. You make up your records, and prepare to begin exclusion of the next category on Monday morning (see the seven-day exclusion trial, page 141).

Now we must get clear what is meant by a category, because unambiguous results depend on getting this right. Suppose that, suspecting your child cannot tolerate milk very well, you exclude it for a one-week trial, and that on Saturday you have nothing to show for your efforts. Your problem could be that other items with the same kind of irritant quality have still been consumed. In fact, most children react not to milk as such but to the kind of protein to be found in cows and in anything they produce. So cheese, butter, cream, yoghurt, beef, beef stock cubes, beef extract, beef suet and milk chocolate must all be included with milk in one category. So must any margarine or package food mix whose recipe contains casein, whey powder, dried skimmed milk, milk solids or any other processed milk product. The category is really 'cow protein', not milk at all.

The same principle applies to every kind of food and chemical. Plants have relationships with each other which are only obvious to gardeners. Potatoes and tomatoes, for instance, are both members of the botanical family 'Solanaceae'. If you suspect potatoes, you should exclude with them all the members of that plant family as one category. And chemicals behave in exactly the same way. If you suspect that orange-coloured fruit squashes may be upsetting your child, you may be tempted to exclude tartrazine alone. It would be better to avoid as many as possible of the other azodyes as well, since they closely resemble tartrazine and will compound its effect. You may widen your net even further to include all the dye chemicals derived from coal tar, whose resemblances are more distant but still significant.

The point of going to all this trouble is that you will get much clearer results, on which you can rely more completely, if you succeed in isolating each category of substances accurately and round up all their members for exclusion during the same week. That leaves no minority to deputize for all the others during the exclusion period, so that on Friday evening they come as a complete surprise. It really is worth learning enough about the subject to ensure you achieve this.

Kinship Amongst Irritants

All living things have been classified over the centuries by botanists and zoologists according to their biological resemblances and differences. Generally speaking, those within the same 'family' are similar enough to arouse the same intolerant response in a sensitive individual. Fishes are a major exception; amongst them it is safer to put 'sub-orders' in one category. You need not worry about that, as it is taken care of already in the food families table at the end of this chapter.

Variety in our food ensures that we avoid dependence on too narrow a range of families, a protection that has been severely eroded in recent decades (chapter seven). The table (pp. 142–51) lists a wide range of these, many of which will be unfamiliar. They are included to give you safe alternatives to try, while avoiding foods from any common families which prove to include items irritant to your child.

Your doctor, health visitor or the dietician based at your local clinic or hospital will be willing to help you plan a satisfactory programme of families to exclude, and suggest alternatives which you can substitute to balance your child's diet during the trial. Some of these will be quite a novelty, and may introduce you to the choices displayed at ethnic food shops. This produce is often much less intensively sprayed than that from Europe and America, and can make life much easier all round for parents facing complicated multiple intolerance problems.

You need not test all the families on the list, of course. Single out those which contain the foods your child relies on most. Any that he particularly craves or consumes heavily are the ones to put at the top of your list. On the other hand, families from which he consumes nothing for whole weeks at a time are unlikely to be causing chronic symptoms.

Chemicals are, unfortunately, a much more complicated subject. They could be properly separated and researched individually at special controlled environment units, but nothing like this exists in England at present. We cannot hope to separate them successfully at home. So we have grouped them much more simply in chemical clans, with an eye for practicality.

Chlorine is closely tied up with water (clan one), which may carry a wide range of other possible irritants in trace amounts. The two are best excluded together in one trial, during which mineral water (in glass bottles) is the best drink. Be sure that no other water is consumed, in other drinks or cooking. If the trial is positive, try boiling tap-water to get rid of the chlorine before consuming it. You can then run a trial excluding chlorine alone. If that gives no positive result, try another trial with mineral water. With precision, and a little luck, you should within three weeks have a clear idea which to exclude in the long term – tap-water, or chlorine.

Tobacco avoidance (clan two) will be harder for the smoking parent than the sensitive child. It is unfortunately not good enough to confine smoking to one room in the house; the fumes get everywhere. The smokers will not themselves notice this, so they need to have it firmly impressed on them. For the exclusion week, they must smoke out of doors or in detached buildings.

If the child benefits and the smoker cannot give it up, an ionizer can be installed in a room designated for smoking, and used whenever tobacco is alight and for some hours afterwards. This helps to destroy smoke which would otherwise hang in the room and eventually leak throughout the house. A candle, burnt for several hours in the room after it has been smoked in works well too; but the sufferer may turn out to be intolerant of the candle!

The smoking-room should be ventilated to the outside, and be separated from the main parts of the house by a lobby, or 'air lock', if possible. All this is a good deal of trouble, but well worthwhile if smoking is important enough to be continued even when you have demonstrated how it harms another family member's health.

A problem about identifying food additives (clan three) is recognizing their names if the E-code number is not given. There are a number of books listed in Further Reading (see page 193) which can help you with that. In our list we give the group name for large classes of chemicals, some or all of which will appear in the full name on the package. Many of these additives are by now familiar; some are not. We have suggested you be wary of ammonium salts, however simple they seem; they add to your body's burden of ammonia, which irritates nervous tissues. Again, we have put the spotlight on all aluminium salts because aluminium is also known to irritate some people. Quite a few additives have no E-code anyway, because they are used in Britain but not approved by the EEC. Those we have listed in alphabetical order at the end.

We meet salicylates (clan four) in aspirin, in some chemical additives in food and in some foodstuffs; so this clan cuts across several others. That is the reason why it must be tested separately, even though this involves items which have already been excluded in other combinations of foods and food additives.

The last three clans are complex and inextricable. None of them can be avoided completely, but with an heroic effort on the part of the whole family for the five-day trial period, a wide range of optional items, such as cosmetics, cleaning fluids, polishes, lighter fuel and dry-cleaning fluids, can be excluded from the indoor home environment. Nothing less than this gives a worthwhile reduction in total exposure from which observations may be possible. Fortunately, if this group of chemicals plays an important part in the symptoms of the intolerant child, improvement is likely to be quick. Then a whole range of long-term issues arises, which are dealt with in chapter fifteen.

Special Tests

If all this is beyond you, or the results are perplexing, you will need to enlist the help of your doctor. If he does not yet feel experienced enough to take matters further, he may be prepared to refer you to a clinical ecologist who can. Some of these will be available through National Health Service clinics and hospitals; but many are forced by circumstance to practise privately. For further advice about your options contact the Environmental Medicine Foundation, whose address is given in chapter fifteen.

Under specialist supervision a dietary trial may be possible in hospital, and for irritant items that cannot be avoided immunotherapy treatment may then be planned. This is done by discovering the exact dilution of a substance which neutralizes its irritant effect when injected under the skin or placed as drops under the tongue. Treatments like these can make a great difference to the life of a severely affected person intolerant of a wide range of substances, as many of the stories in Part Two illustrate.

But most problems will prove to be relatively simple, and readily overcome once they are recognized. That is something you can achieve for yourself, and take justifiable pride in.

The Seven-day Exclusion Trial

Make a list of the food families and chemical clans to which your child is regularly exposed, from the tables (pp. 142–51).

Exclude one family or clan meticulously from Monday morning to Friday afternoon, then reintroduce it as normal from Friday evening.

An obvious reaction during the weekend indicates sensitivity to one or more members of that family/clan.

Go through your list of suspect families and clans one at a time, week by week, recording the results as you go.

Persevere in excluding the ones your child reacts to.

Carry on providing foods from families that do not affect her or him.

You need not meticulously avoid the chemical clans your child does not seem to react severely to, but always take sensible precautions against careless consumption of toxic chemicals, especially garden pesticides.

Food Families
Each family constitutes a category for the exclusion trial described in chapter twelve, and forms the basis for constructing the rotation diet in chapter thirteen. Some rarities are included for their value as alternatives, when the usual choices are blocked by intolerance.

Plants
Aizoaceae: New Zealand spinach
Anacardiaceae: cashew, pistachio, mango
Betulaceae: hazelnuts, filberts
Caricaceae: pawpaw
Chenopodiaceae: beetroot, spinach, sugar beet, Swiss chard.
Compositae: artichokes (globe and Jerusalem), camomile, chicory, dandelion, endive, lettuce, safflower, salsify, scorzonera, sunflower, tarragon
Convolvulaceae: sweet potato
Cruciferae: Brussels sprouts, broccoli, cabbage, cauliflower, Chinese cabbage, horseradish, kale, kohl rabi, landcress, mustard, radish, rape, swede, turnip, watercress
Cucurbitaceae: acorn, cantaloup, cucumber, courgette, marrow, melon, pumpkin, squash, water melon
Cycadaceae: sago
Dioscoreaceae: yam
Ebenaceae: persimmon
Ericaceae: bilberry, blueberry, cranberry, huckleberry, sloe
Euphorbiaceae: cassava, tapioca
Fugaceae: sweet chestnut
Fungi: mushrooms, yeast
Gramineae: bamboo sprouts, barley, sweetcorn, millet, oats, rice, rye, sugar cane, wheat
Juglandaceae: walnuts, hicory nut, butter nut
Labiatae: balm, basil, hoarhound, mint, marjoram, oregano, peppermint, spearmint, rosemary, sage, savory, thyme
Laurus: Herb teas
Leguminosae: dry beans, green beans, lentils, liquorice, peas, peanuts
Liliaceae: asparagus, chives, garlic, leek, onion, shallot
Malvaceae: okra
Marantaceae: arrowroot
Moraceae: mulberry, fig, bread fruit
Musaceae: banana, plantain
Murtaceae: guava
Oleaceae: olive, olive oil

Onagraceae: water chestnut
Palmae: coconut, dates
Passifloraceae: passion fruit
Pineapple
Polygonaceae: buckwheat, rhubarb
Portulacaceae: Miner's lettuce
Rosaceae: apple, apricot, blackberry, cherry, loganberry, nectarine,
   peach, pear, plum, prune, raspberry, rosehip, strawberry
Rubiaceae: coffee
Rutaceae: grapefruit, lemon, lime, mandarin, orange, tangerine
Saxifragaceae: gooseberry, black and red currants
Solanaceae: aubergine, cayenne, chillies, ground cherries, paprika,
   pepper, pimento, potato, physalis, tobacco, tomato
Theaceae: Indian tea
Torreya: nutmeg, mace, Brazil nut
Umbelliferae: angelica, anise, caraway, carrots, celeriac, celery,
   coriander, dill, fennel, parsley, parsnips, samphire
Valerianaceae: lamb's lettuce
Vitaceae: grape, vine, peppercorns

Animals

Anatidae: duck, goose, duck eggs, goose eggs
Bovidae: beef, suet, beef stock cubes, Oxo, Bovril, milk chocolate;
   items including dried skimmed milk, casein, whey powder, milk
   solids, cow's milk, butter, cream, yoghurt, cheese
Caprinae: goat's milk, cheese, yoghurt, sheep's milk, cheese, kid,
   lamb, mutton
Carvidae: rooks
Cervidae: venison
Columbidae: pigeon, dove
Crustaceanae: crab, crayfish, lobster, prawn, shrimp
Leporidae: hare, rabbit
Meleagrididae: turkey
Molluscae: abalone, clam, mussel, oyster, scallop, snail, squid
Phasianidae: partridge, pheasant, chicken, hen's egg
Ranidae: edible frog
Scolopacidae: snipe, woodcock
Suidae: pork, ham, bacon, pig's liver, pork sausage
Tetraonidae: grouse, ptarmigan, capercaillie
Turnicidae: quail*, quail's eggs

* Quail in the wild are protected by law. Hunting them or robbing their nests is an
offence.

Fishes
Acipenseridae: sturgeon
Anguilliformes: eel
Clupeidae: herring, anchovy
Cyprinidae: carp
Gadiformes: cod, haddock
Merluccuidae: hake
Mugilidae: mullet
Percoidei: bass, perch, sea bream
Pleuronectidae: plaice, halibut, flounder
Rajidae: skate
Salmoniformes: trout, salmon
Scombridae: tuna, mackerel
Soleidae: sole

Chemical Clans
These are not distinct chemical classes (which would be impractical), but ranges of substances grouped according to the management problems they present. The easiest to deal with are listed first. For comments on each clan see below.

Clan One: Chlorine, and Water
Chlorinated swimming pools (other methods of disinfection are available – ask the attendant or proprietor).

Chlorine occurs in bleaches, fire extinguishers, anaesthetics, disinfectants, dry cleaning fluids, household cleaners. It is used in processing wood pulp for paper manufacture, preparing vegetables for freezing, refinement of edible oils, refinement of sugar, manufacture of dyes, drugs and cellulose acetate.

Tap water may contain chlorine, chlorinated hydrocarbons, fluoride, polyphosphates, salts of iron, lead and copper, organic matter, nitrates, pesticide and herbicide.

Detergent residues cling to unrinsed crockery etc.

Clan Two: Tobacco Fumes
– include nicotine, cotinine, carbon monoxide, cadmium, lead and smoke.
– are likely to contaminate soft furnishings, wallpaper, glass and ceramic surfaces where smoking is habitual, until these items are cleaned.

Clan Three: Food Chemicals

For detailed listing and evaluation of these see *Additives – Your Complete Survival Guide* (Century 1986).

Avoid especially:

Code	Name	Chemical Family
E102	tartrazine	
E104	quinoline yellow	
107	yellow 2G	
E110	sunset yellow FCF, orange yellow S	
E120	cochineal, carminic acid	
E122	carmoisine, azo rubine	
E123	amaranth	
E124	ponceau 4R, cochineal red R	azodyes
E127	erythrosine BS	
128	red 2G	and
E131	patent blue V	
E132	indigo carmine, indigotine	coal tar dyes (except caramel)
133	brilliant blue FCF	
E142	green S lissamine green acid brilliant green	
E150	caramel	
E151	black PN, brilliant black BN	
154	brown FK, food brown, kipper brown.	
155	brown HT	
E173	aluminium	
E174	silver	metals
E175	gold	
E210	benzoic acid	benzoic acid
E211	sodium benzoate	
E212	potassium benzoate	and benzoates
E213	calcium benzoate	

Code	Name	Chemical Family
E214	ethyl 4-hydroxybenzoate, ethyl para-hydroxybenzoate	
E215	sodium salt of E214	
E216	propyl 4-hydroxybenzoate, propyl para-hydroxybenzoate	hydroxybenzoates
E217	sodium salt of E216	
E218	methyl 4-hydroxybenzoate, methyl para-hydroxybenzoate	
E219	sodium salt of E218	
E220	sulphur dioxide	
E221	sodium sulphite	
E222	sodium hydrogen sulphite, sodium bisulphite, acid sodium sulphite	sulphur dioxide
E223	sodium metabisulphite, disodium pyrosulphite	and sulphites
E224	potassium metabisulphite, potassium pyrosulphite	
E226	calcium sulphite	
E227	calcium hydrogen sulphite, calcium bisulphite	
E230	biphenyl, diphenyl	biphenyl, ortho-
E231	2 hydroxy biphenyl, O-phenyl phenol	phenyl phenols
E232	sodium biphenyl-2yl oxide, sodium orthophenyl phenate	
E236	formic acid	formic acid
E237	sodium formate	
E238	calcium formate	and formates
E249	potassium nitrite	
E250	sodium nitrite	nitrites
E251	sodium nitrate, chile saltpetre	and nitrates
E252	potassium nitrate	
E310	propyl gallate, propyl 3,4,5 trihydroxybenzene	gallates
E311	octyl gallate	trihydroxy-
E312	dodecyl gallate, dodecyl 3,4,5 trihydroxybenzene	benzenes

Code	Name	Chemical Family
E320	butylated hydroxyanisole	BHA,BHT
E321	butylated hydroxytoluene	
E325	sodium lactate	
E326	potassium lactate	lactates
E327	calcium lactate	
380	triammonium citrate	
381	ammonium ferric citrate	
E402	potassium alginate	
E403	ammonium alginate	
E404	calcium alginate	alginates
E405	propane-diol alginate, propylene glycol alginate, alginate ester	
430	polyoxyethylene 8 stearate, polyoxyl 8 stearate	polyoxyls
431	polyoxyethylene 40 stearate, polyoxyl 40 stearate	
432	polyoxyethylene sorbitan monolaurate, polysorbate 20, tween 20	
433	polyoxyethylene sorbitan mono-oleate, polysorbate 80, tween 80	
434	polyoxyethylene sorbitan monopalmitate, polysorbate 40, tween 40	polysorbates tweens
435	polyoxyethylene sorbitan monostearate, polysorbate 60, tween 60	
436	polyoxyethylene sorbitan tristearate, polysorbate 65, tween 65	
E477	propylene glycol esters of fatty acids, propane-diol esters of fatty acids	propylene glycol esters
478	lactylated fatty acid esters of glycerol and propane-diol	

Code	Name	Chemical Family
491	sorbitan monostearate	
492	sorbitan tristearate, span 65	sorbitan esters,
493	sorbitan monolaurate, span 20	spans
494	sorbitan mono-oleate, span 80	
495	sorbitan monopalmitate, span 40	
503	hartshorn, ammonium (bi)carbonate	
510	ammonium chloride	
527	ammonium hydroxide	
541	sodium aluminium phosphate basic	
554	aluminium sodium silicate	
556	aluminium calcium silicate, calcium aluminium silicate	
558	bentonite, bentonitum, soap clay	
621	monosodium glutamate	
627	guanosine 5-disodium phosphate, sodium guanylate	'natural' flavour
631	inosine 5-disodium phosphate	enhancers
635	sodium 5-ribonucleotide	
905	mineral hydrocarbons	
–	acesulfame potassium, acesulfame K	
–	2 aminoethanol, mono ethanolamine	
–	ammonium sulphate	
–	aspartame	
–	dichlorodifluoromethane	
–	diethyl ether, solvent ether	
–	ethoxyquin	
–	ethyl acetate	
–	hydrogenated glucose syrup, hydrogenated high maltose glucose syrup	
–	isomalt	
–	poly dextrose	
–	propylene glycol, propane -1,2-diol.	
–	aluminium potassium sulphate, potash alum	
–	thaumatin	
–	xylitol	

Clan Four: Salicylates

Aspirin (Acetylsalicylic acid) and all medicines including it. Food additives which may provoke salicylate intolerance:

E102	tartrazine
E104	quinoline yellow
107	yellow 2G
E110	sunset yellow FCF, orange yellow S
E122	carmoisine, azorubine
E123	amaranth
E124	ponceau 4R, cochineal red R
E127	erythrosine BS
128	red 2G
E131	patent blue V
E132	indigo carmine, indigotine
133	brilliant blue FCF
E142	green S, acid brilliant green, lissamine green
E151	black PN, brilliant black PN
154	brown FK, kipper brown, food brown
155	brown HT

E210
E211 ⎫
E212 ⎬ benzoic acid and benzoates
E213 ⎭

E214
E215
E216 ⎫
E217 ⎬ hydroxybenzoates
E218
E219 ⎭

E310 ⎫
E311 ⎬ gallates, trihydroxybenzoates
E312 ⎭

Foods containing natural salicylates:

almond	gooseberry
apple	grape
apricot	lemon
blackberry	marrow
blackcurrant	peach
cherry	pepper
cider vinegar	plum
cucumber	prune
currant	orange

raisin	sultana
raspberry	tomato
rosehip	tangerine
strawberry	wine vinegar

Clan Five: Fossil Fuel Combustion Products

Gas Fumes – carbon monoxide, hydrocarbons, nitrogen oxides, sulphur oxides

Paraffin Fumes – smoke, terpenes, polycyclic aromatic hydrocarbons

Petrol and diesel fumes – all the above plus formaldehyde, lead, phenol

Food preservatives E220–E227 – those releasing sulphur dioxide, used in a wide range of processed foods and identifiable from their labels

Clan Six: Solvents

The most troublesome are:

alcohols – surgical spirit, methylated spirit, rubbing alcohol/rubefacients, balms, liniments, tinctures, perfumes, after-shave lotions, deodorants, toilet waters, cologne, wines, spirits, beer, cider, perry

amyl alcohol – a very poisonous industrial solvent

butadiene – residues in sponge rubber, latex

ether – dry cleaning fluid, a permitted food additive

ethylene glycol – permanent antifreeze

glycerol, glycerine – food sweetener and preservative, anti-freeze, cosmetics, perfumes, soaps, toothpaste, cologne, toilet water, ink, some glues and cements, medicines and suppositories, confectionery, confectioners' icing, ice cream, cakes and biscuits, handcreams, artificial fabrics

halocarbons – commercially frozen food, decaffeinated coffee, aerosol spray cans including medicated inhalers

isopropyl alcohol – antifreeze, rubbing alcohol, solvents

menthol – perfumes, confections, liqueurs, medicines and lozenges, some cigarettes

phenols – carbolic soap, medicated gargles, herbicides, pesticides, epoxy resins; phenolic resins – bakelite, moulded plastics, laminated boards, thermal insulation panels; synthetic detergents, petrol additives, dyes, photographic solutions, preservative in medications and injections, casing of electric cable and flex, perfumes, deodorants and deodorizers, candles, ink, toilet tissue, lipstick; residues from manufacture of aspirin, nylon, polyurethane, explosives based on picric acid. Naturally present in poison ivy, poison oak, thyme oil and springs arising near rich humus or coal

styrene – polystyrene products, rubbing alcohol, solvents

terpenes – paint fumes; house plant taste, smell; resins and latex of rubber plant; domestic gas coal and oil, motor exhaust fumes; anaesthetic from petrochemical sources; lighter fuel

Clan Seven: Formaldehyde (Formalin)

cavity wall insulation

textiles, and a wide range of processes applied to them, e.g. dyeing, pressing, moth proofing, shrink proofing, fabric softeners and conditioners

foam rubber fillings – mattresses, cushions, carpet backing. Orthopaedic plaster casts.

propellants in spray cans

resin glues and cements

milk; butter, cheese and milk products in UK (as preservative)

paper and newsprint

photography and photographs

antiperspirants

disinfectants, dentifrice antiseptics, mouth washes, nail polishes, toothpastes, soaps, shampoos, hair setting lotions

air deodorant

insecticides and fertilizers (garden and farm)

concrete, plaster, wallboard, wood veneers, plywood, chipboard, blockboard, wood preservative

manufacture of antibiotics, vitamins A and B

traffic fumes, petrol and diesel fumes

Clan Eight: Pesticides

Huge variety – several hundred individuals, in a hundred distinct chemical classes

   Found in – insect sprays
               fly papers
               garden spray chemicals
               carpets and blankets (as preservatives);
               carpet shampoo
               moth balls
               lotions and shampoos for treatment of body lice, head lice, nits
               agricultural sprays and spray drift.

CHAPTER THIRTEEN
Helping Your Self

With luck, you know by now what causes your child's intolerance symptoms, and are ready to set about keeping it from him
for ever. And that may be all you need do to restore him
permanently to perfect health and vigour.

Sometimes, however, the problems turn out to be a lot more
complicated than that. If you saw clear reactions to seven food
families and suspect trouble with a couple of chemical clans as
well, what are you to do? Avoiding all of these will be difficult or
impossible, and the attempt will turn your child into a kind of
environmental cripple. Long-term reliance on medical help is
no better, however effective it may be. No one should be
launched on life with the apprehensive, dependent and negative
outlook these produce, if you can help it.

You have seen in Part One how things have gone wrong,
which is the basic knowledge you need to put many of them
right. People all over the world are researching how to do this,
and you can take advantage of what they are discovering. For
instance, Dr William Rea and his colleagues at the Environmental Health Centre in Dallas, Texas have established a
controlled experimental environment in which people can be
carefully excluded from their daily doses of chemical irritants,
and have been able to study and report on the effects of a
great many different kinds. The really good news is that, in
a clean environment, your body can in a matter of weeks
get rid of the chemical residues it contains. What is more,
large and progressive improvements in general health and
mental function come about quite quickly as the residues are
excreted.

This beautifully demonstrated healing process is not confined to artificial environments and experimental conditions. It

is going on all the time, in just the same way, in your ordinary every-day life. However severely your child continues to be exposed to irritants, his body goes on faithfully trying to cope with them. Many of the things we do get in the way of this and spoil the result. By setting that right, we can greatly improve the healing power available to us for coping with chemicals and every other kind of stress. That reduces their amounts and effects in our bodies, even if exposure carries on just as before. And it need not, because intelligent changes in our way of life can reduce that too.

This gives you plenty of scope for a positive long-term campaign for reconstructing health. And what is good enough for the worst-affected members of the family, will do very well for the rest of you. If everyone takes part in the campaign it is easier to keep it up, and within a few months you will all be glad you did.

This chapter and the next spell out what you can do. We start with the most obvious and important topic – food.

Feeding Your Health

Nutritional science has been in the doldrums since biochemists took it over, and is now usually presented as a boring list of technicalities and tables. We shall go on as we started, and think biologically. That means no hard and fast rules, but a number of general principles. Honour these, and food will soon begin to fascinate you. Rediscovery of the lost pleasures of real meals will be a hobby, not a chore.

Make a start by deliberately eating something raw and still alive, at the beginning of every meal if you can. Freshly grated carrot is an easy and popular choice, plain or dressed with a simple sauce of oil and vinegar or lemon juice. Sprouted seedlings are another option, easy to grow and something the children can do. Use several varieties, and eat them young and tender (see below). In the main growing season, of course, you can choose from a much wider range of possibilities, but you need to be sure that they are freshly picked to be eaten still alive, as well as raw.

SPROUTING SEEDS TO EAT LIVE

You will need four clean glass jars, for example 1 lb coffee size or 1½ lb Kilner jars. Replace lids by a layer of muslin or gauze secured by a rubber band or elastic.

Put beans or seeds to 1/5th of the volume of the jar. Cover with water overnight. Cover with muslin and secure with band. It may not be necessary to remove this until you take out sprouts to eat.

In the morning drain through the muslin. Rinse in cold water and drain again, removing as much moisture as possible to permit the air to circulate during the day.

Rinse and drain in a similar fashion twice a day. For small seeds (cress), emphasize the drainage; large seeds (beans) need more frequent moistening.

Once the sprouts are at least as long as the beans or the seeds from which they come, they are ready to eat. They may be eaten whole and raw without further washing.

It takes two to three days for each crop to grow to this stage, after which they are edible for several days more if rinsing and draining is continued. As one jar comes ready another should be started to maintain a continuous supply.

This has two great advantages. Live food still has its own enzyme systems intact and running, and will in your stomach start the process of its own digestion. You take full advantage of this, once you give yourself the chance. This is not an accidental coincidence: remember the million-year-long evolutionary tradition you are drawing on. Your digestive juices team up with the plant's metabolism to produce a reliable, efficient and comfortable result, with no mischievous or useless by-products and fermentations. Try it for yourself: sprouted bean-shoots will not make you windy, like baked beans often do.

But the best news is yet to come. We know from experiments done many years ago by Dr Max Bircher-Benner, inventor of the famous muesli, that cooked food of any kind provokes in the skin of the intestine a reaction like the early stages of inflammation. It seems to act as a general irritant, part no doubt of the mechanism of all food intolerances. But Dr Bircher-Benner went on to show that live food does not provoke this irritable reaction; the bowel lining seems completely at ease with it. What is more, cooked food which follows the live course in the same meal, does not produce the usual irritation either! The live food sets up some kind of protective understanding, which very much reduces friction between your body and everything the meal contains. What an obvious advantage this is to any person sensitive to food, however long the list of risky items!

This discovery may encourage you to go out into the garden and dig for victory again. This time, the enemy is disease. Anything produced straightforwardly from honest soil, without packaged chemistry, is worth its weight in gold. The neglected gardens of Europe are amongst the best places to start – they alone, it seems, have escaped half a century of chemical abuse. Think of this as you labour to produce a seed bed, and return repeatedly to weed it. Humble and thankless though the activity may seem, it has fundamental worth. It may be the cleanest thing you ever do. And if you should find in it a real escape to tranquil reality from the frightening pace of modern life, you are simply joining thousands of Americans and Europeans all making that rediscovery.

Our tip is simply, do not rush at it. If gardening is new to you, get one of the gardening books listed in Further Reading and read it carefully first. Budget a certain period of time weekly to the garden, and do not fret about the weather; be flexible enough to postpone your session a day or two, until it improves. We should be surprised if you are not amply rewarded, with worthwhile crops and a wholesome new outlook on life.

Softly, Softly
These simple starts will encourage you in more tedious efforts – gradually excluding things from your diet that you probably like, but should try to avoid in future. There is a short list of items everyone should cut out or drastically cut down, because they always cause trouble in some form sooner or later. They are refined flour, sugar, stimulants and a range of about sixty irritant chemical food additives.

We listed earlier (p. 52) the additives in white flour, and the nourishment taken out. Do not imagine that ordinary brown flour is any better: it is usually just coloured. 'Granary' type flour is often coloured too, with some wheat sprouts added to give it texture. Even 'wheatmeal' is usually refined, at least in part. And 'made with wholemeal flour' seems calculated to mislead; a proportion of wholemeal can quite legally be added to a majority of white flour, and still justify that description. The only terms which guarantee the article are 'Wholemeal' alone, and 'Wholemeal Bread'. And once you have found a brand, you will taste straight away how much more body and flavour it has. You will find that less of it goes much further; jaws and intestines really perk up and take notice!

Sugar presents a greater problem for many adults, but is easier to stop for children than many people realize. The key to success is replacing it with equally enticing alternatives. Chopped dates, raisins, fresh fruit and figs are just some of the options. But do not be misled into thinking that 'brown' improves on 'white', for sugar any more than for flour. Except for raw cane molasses and genuine raw whole Muscovado and Barbados sugar, some degree of refinement is involved. The safest straight alternative sweeteners are malt and honey, but even these need to be used in small quantities. At least they taste of something and are nutritious, as well as being sweet. Many children would find molasses too bitter as a sweetener, but you could use it successfully as an alternative in cooking.

Stimulating

Look at all the most popular non-alcoholic drinks in the developed world and you will find they have something in common. They perk us up, and arouse us to do things. This is the perfectly genuine chemical effect of the xanthines, a class of substances these drinks all contain. They are naturally present in certain tropical nut kernels, and the leaves and pods of certain Asian herbs. You will recognize them better by their separate names – caffeine, theophylline and theobromine.

Coffee, tea, cocoa (and chocolate) and cola owe their popularity to this stimulant effect. Used occasionally and in moderation, with intervals long enough for proper recovery to normal, there is no harm in them. But if you regularly use them to keep you going through every day, you are on dangerous ground. They can cause anxiety, trembling, irritability, sleeplessness, headache and mental confusion. Muscles are restless and over-active. The heart easily races away in palpitations, and some irregular beats and butterfly feelings begin to intrude. Limbs get cold because less blood flows there.

Not many children drink tea and coffee, or get it at full strength. But thousands drink cola and eat chocolate every day. If they show signs of strong attachment when it is withheld, you have identified an important contributor to their problems. Often, it is the only thing you need to change. It will still be difficult. You may have to endure a week or two with the child craving for it, and showing in exaggerated form any or all of the undesirable features you want to be rid of. Or he may just be drowsy, listless and apathetic. Stick with it, and he could well emerge a new child. Without any doubt, getting him off it is worthwhile.

When you have seen what it was doing to him, start on yourselves. I once heard a story from the executive suite of a prominent local industry. Its members were so often meeting each other around the coffee machine that they decided to get rid of it. For the next three weeks several of them felt intolerably drowsy, and kept needing to go out for air. One or two were found slumped across their desks in the afternoon, sleeping peacefully. A benevolent board overlooked all this, and it soon passed. I hope they were amply rewarded by the subsequent wisdom and calm clear-headedness of their senior employees!

Whatever are you going to drink instead? You can try decaffeinated coffee, but do not offer it to a child with intolerance problems. It can still contain residues of the halocarbon solvents used to process it. And not all the caffeine is removed anyway. Go for a complete change.

Everything else will seem tame, of course; you must expect that. Try rooibosch, or some of the wide range of other herbs available for making tea. Various coffee substitutes are made with roasted barley and chicory. Hot diluted fruit juice suits many people. Lemon barley water is very refreshing and nutritious, particularly in convalescence; but it must be made to an old-fashioned recipe (see below). And you must have something you can do with all the left-over pearl barley!

HOME-MADE LEMON BARLEY WATER

(Mrs Beaton's recipe – much more effective than commercial cordial)

Take 2 oz pearl barley; cover with cold water in a pan, bring to boil.
Strain off the water and discard it.
Add back 1 pint of freshly boiling water, stir, and leave to steep and cool a little; ½–1 minute is usually enough.
Strain off the barley water. Add lemon juice, and honey if liked, to taste.

Vegetable extracts make a good hot drink similar to beef tea. And a tablespoon each of good honey and organic cider vinegar, diluted to taste with hot water, is an excellent natural trace mineral supplement. For cocoa fans a successful alternative is carob, though you will have to exercise a little patience and skill to persuade it to emulsify into a smooth creamy drink. You have to do for yourself the work of the additives in drinking chocolate.

Those Additives

If by this stage your problems are not completely solved, and you have not yet tackled the formidable list of chemicals now permitted as food additives, you can postpone it no longer. It is not really quite as difficult as it seems. Ever since the E-code list was published people have been studying it from the consumer's point of view. See Further Reading (page 193) for two versions, in an inexpensive pamphlet and a full-length paperback, to help you make sense of it. Both classify the additives from various points of view, according to your interest.

With intolerance particularly in mind, we listed the most important irritants on page 145, as chemical clan three. If you can make some progress in checking the amount of these your child consumes, you have done something worthwhile.

Make a start by setting out on the table the loot from a normal main shopping expedition. Go over each item with a red felt-tip pen, marking every ingredient named on the label which crops up in the list. You will quickly see which are the most common, and realize how often they are used. Sort the packages into groups according to how many suspect ingredients they contain, and make a black list of the worst group.

Next time you shop, allow time to browse around the shelves looking for alternatives to your black-listed items. Some categories, like ketchup, are easy to improve; but the quantity you get for your money will be much reduced. Go for quality, and discipline everyone at home to eat less of it. They may even start tasting the tomatoes!

In other categories you may be forced to change your store. Greengrocers will be pleased to see you back, and good wholefood shop proprietors light up when people come in wanting to give their wares the third degree. They can usually re-introduce you to cooking methods old and new, with a wide range of inexpensive paperbacks. Play the field. Little by little, as your stocks need replacing, you can go on trying new things. Be in no hurry, your family can use a little time as well as you. Rapid changes only unsettle digestion and arouse protests you do not need. You will find yourselves exploring a wonderland of fruits, vegetables, herbs and cereals you never knew existed. With these to replace it, less meat will not strike you as any hardship. You can get the children to mix rolled grains bought separately, to make a muesli costing half the price of bag mixtures. You can go one better, and actually eat it with fresh raw fruit as you are meant to. And give someone the job of soaking overnight the dried fruits for the morning. They taste much nicer, go further, and lose a lot of the sulphur dioxide used to preserve most of them.

Within a couple of months you will have come out on top. After that, be careful not to bore your friends with what you have discovered. Wait, and let the effects on your family sink in. When neighbours want to know, they will ask you soon enough.

Going Organic

This is now but a short step, and in some cases an important one. We saw in chapter two how cereal crops are often sprayed right up to the dry spell before harvest. That means the grains inevitably carry residues of weed killer and pesticide into store

BRITISH SYMBOLS
OF ORGANIC QUALITY

The four organizations listed below all maintain standards which guarantee full organic crop production – on biologically composted soil, without pesticides of any kind (herbicides, insecticides or fungicides). The symbols illustrated are their Registered Trade Marks.

Any of them will be pleased to receive enquiries for the names of local suppliers able to offer produce grown to satisfy their standard.

Biodynamic Agriculture Association
Woodman Lane
Clent
Stourbridge
West Midlands DY9 9PX

Organic Farmers & Growers Ltd.
   9, Station Approach
   Needham Market
   Ipswich
   Suffolk IP6 8AT
This organization also maintains
standards for various intermediate
grades, not fully organic. Consult
them for details.

OFG

O.F. & G. Grade I

The Soil Association Ltd.
86 Colston Street
Bristol BS1 5BB

The Farm Verified Information Centre
   86 Easton Street
   High Wycombe
   Buckinghamshire HP11 1NB

with them. If the cereal is to be refined the skin will be removed complete with most of these spray residues, so that some protection from them is obtained by accident. Ironically, by eating our grains whole and unrefined we get these residues back, a new hazard to spoil otherwise valuable benefits.

In a tiny minority of the agricultural fraternity, this problem and others like it are well-recognized. Four separate but cooperating organizations are now busy maintaining and promoting standards of crop production which guarantee freedom from spray residues by the only legitimate means – not using them. They go further, and undertake to fertilize the soil in biologically acceptable ways, so as to restore its richness and put heart back in the crops it grows.

When you are buying flour and whole-grain cereals in particular, look on the package or display for one of the organic symbols illustrated here. If you cannot find them, ask; or enquire further afield. Nothing will encourage this thoroughly good trend more certainly than the desire of many customers, clearly expressed.

Superchargers

A diet like this will serve most people splendidly all their lives. But it may not be quite enough to get your child out of any trouble he is already in. He may need more intensive nourishment to get well, so that he can stay well in future on straightforward food.

Researchers all round the world are paying a lot of attention to elemental nutrients. These are concentrated formulations, of single substances or combinations, designed to fortify diets for particular purposes. Most doctors regard them as an expensive self-indulgence, and so they are when used to paper over the wrinkles of an unhealthy way of life. But sometimes, for a few weeks or months, they can be most helpful.

All the ones we recommend to you are well-established key ingredients in important body mechanisms. If these processes are deranged or lying idle, it may take strong persuasion to get them working again. Supplements can do this, like a start-up system in an idle industry. Putting in plenty of all the tools required and a fair sprinkling of skilled workers, soon gets it back in business. And once all the machinery is well-oiled and running smoothly, the shock troops can be withdrawn.

The best all-round cleanser we have is vitamin C, or ascorbic acid. This improves the entry of nutrient minerals, and dislodges the toxic ones. It defuses free radicals, the highly reactive and disorderly fragments of chemistry which are broken free by chemicals, cosmic rays and irradiation from nuclear fallout like that from the Chernobyl accident. Your white blood cells need loads of this vitamin when they are actively engaged in defending or repairing parts of you. Your skin is cemented together with it; no weaknesses in your outer defences against chemicals and infection can be repaired without it.

For routine purposes, most people need about 200 mgm daily of this vitamin. But you can often make use of much more, depending on how much mess you have to clean up. If you are constantly under stress, infected, smoking or taking the contraceptive pill you can use 3000 mgm and still look for more. A simple test on your urine with silver nitrate solution can demonstrate that, but it would be cheaper just to get on with consuming more. A crystalline powder of pure vitamin C provides you with 750 mgm per quarter level teaspoonful. We suggest you give a child in bad trouble with intolerances this amount in dilute fruit juice, three times a day for at least a month before gradually cutting it back to a third of that.

At the same time, add to one daily drink a teaspoon of a solution of zinc sulphate, prepared to the recipe given below. That gives a full adult supplement of 15 mgm zinc daily, which a deficient child can take advantage of. If it tastes bitter to him he does not need so much – zinc is necessary for a normal sense of taste. In any case, do not continue it beyond the first month or two, repeating it afterwards if he seems to need it.

A child with skin problems or hair and fingernails in bad condition, with weak defences against infection, or with a generally pathetic or hyperactive nature may perk up very quickly with zinc. But to take full advantage of it you need to make sure he also has plenty of vitamin B6 (pyridoxine) available, as these two work together. A child who can describe his dreams in detail at breakfast-time is unlikely to be deficient; for anyone else, a supplement is inexpensive and does no harm.

Fresh wheatgerm, one or two dessertspoons daily, will be sufficient for a child who can have wheat; six vitamin yeast tablets will do if fungi are permitted. Otherwise, several manufacturers provide yeast-free B complex tablets. The names of those we have found reliable are given at the end of this chapter.

Yeast is also a good source of selenium, a mineral we now know to be important in immune function. If yeast is out, the same manufacturer provides more expensive tablets of selenium, or the chemist can make you up a solution containing sodium selenite 200 microgm (0.2 mgm) in 5 ml, to give him one teaspoon daily for a month. That needs vitamin E to work with, a little of which is found in fresh vegetable oils. For this purpose get capsules containing 100 international units of vitamin E, and give him the contents of one whisked in some juice, with the vitamin C and zinc, half an hour before his main meal. (If he can manage to swallow the capsules whole, so much the better.)

ZINC AND SELENIUM
SUPPLEMENT SOLUTION

Hydrous zinc sulphate BP	2.0 gm
Sodium selenite	6.6 mgm
Water	150 ml

Dose   one medicinal teaspoon (5 ml) daily,
child under five, scant ½ teaspoon
giving   zinc 15 mg
selenium 100 microgm

Oils bring us to the next item in this supercharging package – a selection of the essential fatty acids we discussed in chapter six. Evening primrose oil has been widely available for several years now, but the newcomer blackcurrant seed oil is more comprehensive and gives good value for money. We suggest adding the contents of one 500 mg capsule to the main daily dose. Or you can massage it into a patch of skin, which will absorb it. If that patch begins to improve, you can have high hopes for the rest of his skin eventually.

Lastly, you must make sure that the other trace nutrient minerals are there in sufficient supply. The cheapest way is

SUPERCHARGER PROGRAMME
Cost for 30 days – £10–£12

YOU WILL NEED
1 Zinc and selenium supplement solution   150 ml
2 Vitamin C powder (e.g. Boots)   60 gm
3 Your choice of vitamin B:
    either wheatgerm   500 gm
    or   yeast tablets   200
    or   yeast free B complex
        (e.g. Nature's Own)   30
4 *either* organic cider vinegar
        (e.g. Aspall's)   1 litre
        good local honey   2 lb
    or   multimineral (e.g. Nature's Own) 60

PROGRAMME
Morning:
    Diluted pure fruit juice containing quarter-level
        teaspoonful of Vitamin C powder
        (e.g. Boots)
Thirty minutes before main meal:
    Diluted pure fruit juice containing quarter-level
        teaspoon vitamin C
    one teaspoon zinc & selenium supplement solution
    *One capsule vitamin E 100 international units
    *One capsule blackcurrant oil 500 mgm (Lanes)
Daily:
1 *either* Drinks as desired of:
          cider vinegar 1 dessertspoon
          local honey   1 dessertspoon
          hot water to taste
    or multimineral (Nature's Own) 1–2 tablets daily
2 *either* To mashed potato, sauces, or cereal, add
        1–2 dessertspoonsful fresh wheatgerm
    or Take two vitamin yeast tablets three times daily
    or Take one yeast-free B complex

---

* If gelatin cannot be tolerated, or capsules not swallowed, cut both capsules in half lengthwise and drop them in the juice. Whisk briefly, and strain out the capsule fragments.

through seafood if permitted, and cider vinegar and honey (see supercharger programme, page 164) otherwise. If exclusions are a real headache you can rotate these two items, as explained below. Alternatively, a good balanced comprehensive formulation of trace minerals is obtainable from the manufacturers listed. Two per day of the best will be your most expensive item. Even so, the entire package of one month's supercharging need not cost more than about £10.

The details are summarized in the chart here for easy reference. Stick to them exactly, and be content with just one month. You can have too much of a good thing.

Water Fit to Drink

In a concerted effort like this, remember how often in every day you reach for the tap or kettle. Especially if chlorine, insecticides or trace heavy metals look like being problems for your child, take the trouble to filter or replace it.

Local springs are tempting, but prey to all sorts of problems that it may be costly to investigate. You would be better off for short-term efforts to buy bottled spring water from a reputable deep source well away from intensive agriculture. The content of most of these is well known, and is often summarized on the label. Write to the distributors for further details if that does not answer your particular questions.

Unfortunately, many of these are now distributed in 1½-litre plastic bottles. Gaseous residues from the plastic can and do dissolve into the water, and can still be a nuisance for a very intolerant child. You will then have to consider the more expensive brands in glass 1-litre bottles, like Malvern and Perrier.

For long-term changes, consider an alternative approach. Filtration equipment is now being developed to suit various particular needs and pockets. Listed below are several commercially available options. The performance of each varies, but even those which just remove suspended particles make a lot of difference to drinking quality. All of them remove chlorine.

WATER FILTERS

Manufacturer	Equipment Type	Performance	Cost Initial	Annual
Fileder Systems Ltd, 50 Old Road, Wateringbury, Kent ME18 5PL	'Ametek': plumbed in	chlorine, organic taints, sediment	£215	£5
	tap attached	Up to 90% chlorine, fluoride, nitrate, sodium – *but* also iron, calcium, magnesium, potassium.	£75	£40
Amsoil Whitehaven Products, Leigh Road, Bradford-on-Avon, Wiltshire BA15 2RS	'Aquabrite' plumbed in below sink.	organics, mercury – high proportions. *not* lead. fluoride type available.	£150+	£25–30
Brita (UK) Ltd, Yssel House, Queen's Road, Hersham, Surrey KT12 5NE	'Brita', jug.	lead – 100% copper, organics, chlorine – 90%. cadmium, mercury – up to 70%. *not* fluoride.	£12	£12–£24
Fairey Industrial Ceramics Ltd., Filleybrooks, Stone, Staffs. ST15 0PU	'British Berkefeld', Super Sterasyl NFU and NHNS both above sink, plumbed in.	bacteria, chlorine, organics, pesticides, heavy metals, fluoride	£97 £117	£13 £18
La Source de Vie, PO Box 66, Chichester, West Sussex, PO18 9HH	'Mayrei 2000' attaches to water tap.	mercury, copper, & chlorine 90% lead, cadmium – 50–70% sediment – 80%; nitrate – 20%	£10 or so, monthly	
Fospur Ltd., Dudley Rd., Kingswinford, DY6 8XF	'Safari' AB1/C or AB1/F (incl. tap) Both are plumbed in below sink.	suspended matter, chlorine, detergents, organic taints, asbestos.	£99 £160	£40 £40

Merry-go-round

Even if by faithfully following all the suggestions in this chapter you cannot resolve all your child's intolerance problems, do not despair. It may still not be necessary to avoid them all, nor to

SETTING UP A ROTATION DIET

1 Cross out of the list of food families (pp. 168–170) those of which your child is seriously intolerant.

2 Select from the remainder all the food families available and acceptable to you.

3 Copy out each food family three times, on three separate sheets of paper, and stack them in order. *numerical?*

4 Lay out on the table seven titles, one for each day of the week.

5 Take the three sheets for the first food family and set out one on each of three consecutive days of the week, by putting them under those titles on your table. Now do the same for every other food family in turn, until all the sheets are used up. Rearrange them to suit your habits and convenience, but keep each family on three consecutive days. That ensures an adequate rest between portions, but gives you a chance to use up left-overs and try different recipes.

Remember you can overlap from the end of the week, back to the beginning.

6 See that enough variety of choice is available on each day, by checking that the families listed spread well across the different kinds of food – vegetables, fruits, carbohydrates, nuts, meat, fish, fats and oils, dairy produce, drinks and sweeteners.

7 Bear in mind that milk, soya and lemon juice (citrus family) can be used interchangeably. Soya milk can be substituted in drinks, puddings and cereals on non-milk days, and lemon juice is an alternative for either in teas.

The same principle applies to sweeteners, fats and oils.

8 When you are satisfied with the daily options make out permanent lists, one for each day. It is helpful to classify the list in kinds of food (step 6 above) to form a clear summary on which to base your shopping expedition.

punctuate his life with 'don't' or 'may not' frequently repeated. Try to set up a rotation diet instead, which keeps to within three consecutive days of any week every food he has. This minimizes the general level of dietary irritation, so that he can better tolerate small quantities of inevitable daily exposures of other kinds. You can thoroughly satisfy his desire for anything during a three-day span, and will not often be caught with unusable left-overs. It is highly practical, and helped many of the children mentioned in Part Two.

The rotation diet (see previous page) is based on the list of food families given below. Here, they are arranged a little differently and numbered for reference. Though it may look complicated, each step is simple, and you can revise your choices whenever you need to. But within one carefully chosen set of daily menus there are enough different choices to keep most people going for years.

Food Families for Diet Planning

Note: this list differs slightly from that on pages 142–144, to make it more practical in daily use. The Latin names are replaced by numbers.

1	lemon	8	chicken
	orange		turkey
	grapefruit		duck
	lime		goose
	tangerine		pigeon
2	banana		quail
	plantain		pheasant
3	arrowroot		eggs
4	coconut/oil	9	grapes
	date		raisins
	date sugar	10	pineapple
5	parsley	11	strawberry
	carrots		raspberry
	parsnips		blackberry
	celery		loganberry
	anise		rosehip
	caraway	12	melon
6	pepper		watermelon
7	nutmeg		cucumber
	mace		courgettes
	brazil nut		cantaloup

pumpkin
acorns
squash
13 sugar beet
spinach
swiss chard
14 pea
dry beans
green beans
soy beans/milk/oil/flour
lentils
liquorice
peanuts/oil
15 cashew nuts
pistachio
mango
16 hazelnuts
filberts
17 swine (pork)
18 abalone
snail
squid
clam
mussel
oyster
scallop
19 prawn
shrimp
lobster
crab
crayfish
20 apple
pear
21 figs
mulberry
breadfruit
22 olive/oil
23 currant
gooseberry
24 buckwheat/flour/grain
25 lettuce
chicory
artichoke
dandelion

sunflower/oil//marg
tarragon
26 potato/potato flour
tomato
aubergine
peppers
paprika
cayenne
ground cherries
pimento
chillies
27 onion
garlic
leek
asparagus
chives
28 mint
sage
thyme
oregano
basil
marjoram
savory
rosemary
hoarhound
peppermint
spearmint
29 walnut
hickory nut
butter nut
30 chestnut
31a salt water fish:
herring
mackerel
sprats
anchovy
sea bass
sea trout
tuna
swordfish
flounder
sole
31b cod
haddock

whiting
31c plaice
32 fresh water fish:
   sturgeon
   salmon
   whitefish
   bass
   perch
33 plum
   cherry
   peach
   apricot
   nectarine
   almond oil
   wild cherry
34 blueberry
   huckleberry
   cranberry
   sloes
35 watercress
   Brussels sprouts
   turnip
   cabbage
   broccoli
   cauliflower
   horseradish
   mustard
   radish
   swede
36 avocado oil
   cinnamon
   bay leaf
37 wheat
   corn/oil/rice/flour
   oats
   barley
   rye
   cane
   millet
   bamboo sprouts
38 mushrooms

yeast
39 cow's milk
   butter
   cheese
   yoghurt
   beef
40 lamb, goat's milk, cheese,
      yoghurt
41 rabbit
   hare
42 rooibosch tea
43 breakfast herb tea
44 decaffeinated coffee
45 dandelion coffee
46 carob
47 camomile tea
48 peppermint tea
49 rosehip tea
50 chocolate
51 okra
52 sprouting seeds
53 allspice
   cloves
54 Tomor margarine
55 Golden Rose margarine
      (Kosher)
56 sago
57 sweeteners
   pure honey
   pure fructose
   pure glucose
   cane sugar – see 37
   corn syrup – see 37
   beet sugar – see 13
   date sugar – see 4
   oils/fats:
   from foods, meats as listed
   milk:
   goat's milk – see 40
   soya milk – see 14

CHAPTER FOURTEEN

Spring Clean

In the previous chapter we accentuate the positive. Now we must eliminate the negative – an equally constructive exercise, but more formidable.

Do not be dismayed by what seems at face value an impossible list of tasks. Actually you only need to tackle them one at a time. People who try are amazed how fast in practice they can adapt to ideas like this. It does not make you hypochondriacal or obsessive, if you were not so already; though it obviously heightens your awareness of the real state of things. That may motivate you to more strategic objectives, which we discuss in our final chapter.

Every one of the manoeuvres we mention now will lighten the burden of chemical stress on every member of your household. Anyone showing intolerance already will gradually regain their rightful resilience. The rest of you will be aware of a general increase in your vitality, reduced irritability, far greater ease in tackling physical and mental work, and a greater appetite for life in general.

Anyone who perseveres stands to experience a priceless sense of youth regained. It is just as if you bob up to the surface of a sea of troubles, exchanging buoyancy and lightness for the slow heaviness and frustration you now take for normal life. If you cannot remember when you last felt irrepressibly hopeful, or sang for sheer joy and pleasure, then it is time you demolished the obstacles.

We start with things that everyone can do, as soon as you decide to start. Then little by little, as the opportunity occurs, you can tackle the more ambitious tasks we mention further on.

Coming up for Air

Most people only ever use a fraction of their lung capacity. Life does not call on us to toil like people used to, and an occasional burst of sporting exercise is not good enough instead. But if you are prepared to get a few deep breaths into the furthest recesses of your lungs each day, then you will keep all the capacity for ventilation you were born with.

You can do this without any change of routine, by remembering to take four or five long, slow, deep breaths in through your nose, at the same times every day. Waiting for the bus, washing up and queueing in shops are just three opportunities. It works best if you are not tightly belted at the waist, so that your stomach can swell out as you breathe in. This feels very unnatural at first; we are used to breathing with our ribs alone, and feel ugly when we let our abdomens stick out. But no one can see you doing it under normal clothing – you are not asked to make a public scene! And the increased capacity of each breath taken with your belly will surprise you.

Hold each breath in for a slow count of five, and let it out gently for another five. Breathe in again immediately, taking five to do it. Four such breaths, taking only a minute, will feel like hard work at first. But they will elevate and calm your mood, renew your energy and clear your mind as well. Make it a habit, and the effect accumulates day by day.

Getting your lung capacity back is one thing. Now try using it. To keep your heart and lungs efficient takes much less effort than you think. Ten minutes' exercise three times a week is enough to gradually restore reasonable fitness, and you do not even then have to look energetic! Just choose a gentle form of exercise you can enjoy, that fits well into your daily life. Walking briskly is quite enough. Cycling to the shops may be more practical; the saving in short car journeys soon pays for the bicycle. Children love biking too, so a regular family ride is easy to establish and great fun for everyone.

You need only work hard enough to make your breathing noticeably heavier. Conversation should be difficult, coming in short utterances with each breath. If you go in for counting your pulse, it should not rise above 120–130 beats per minute. Nothing need hurt. Yet regular practice of this habit will increase your staying power, and create a decent turnover of air and energy. That gets the blood to where the chemicals are, and gives them every chance of blowing away in your breath.

If the air round where you live always seems polluted, get up early before traffic and factories get started. Morning air has a specially refreshing quality which will reward your effort. And if you find yourself ready for bed earlier at night, so much the better. Get back your early morning gladness, and you have really scored!

Rinsing Off

When your heart and lungs are working well for you, think of your skin. It does not sweat only when you are exercising. Make the experiment of wearing a plastic bag inside your shoe one evening; the moisture that collects will be appreciable by bedtime. Or put a plastic sheet under your mattress, and inspect it after a few days. The dampness has come from your normal night-time perspiration!

We wash more often than we exercise, so an efficient toilet routine offers just as much benefit as harder work. Skin clogged with chemicals is an opportunity lost. Before you dress, wipe all over with a warm damp facecloth, then brush vigorously with a clothes brush or loofa. It should not scratch, but raise a uniform pink reaction in the skin. It gets rid of the scales of unwanted skin that are ready to be shed, and invigorates the nervous control of all your automatic functions, not just the skin. If you doubt this, try it. You will feel larger than life for an hour or two afterwards.

At the end of an active day your skin will be clammy with dried-out perspiration, and a shower will be most invigorating. After a hot relaxing rinse, turn the heat off. Half a minute of cold water on a hot body closes the pores, squeezing out the last little bit of sweat you would have kept there. It sets you up for the evening, and makes sound sleep more assured. If you have no shower, use a quick all-over wash from a shallow bath or basinful of water. Soaking daily in a deep bath is expensive and unnecessary.

If chlorine seems to irritate you, neutralize it with a teaspoon of photographer's hypo (sodium thiosulphate) in each bath, and run it with the window open. A less sensitive member of your household may perhaps do this for you, if you are badly affected. A few grains of hypo is sufficient in a basin of washing up: you can get it from photographic suppliers or chemists.

Do not then undo your efforts with chemical toiletries! Only use soap if you are dirty; water alone is quite sufficient otherwise. Non-greasy dirt comes off with a little cold cream. Even long hair washes well with cold cream or soap, if you are prepared to give it the time. Water alone will do sometimes, especially if your hair is short and dry rather than greasy. Soap and its additives are really very harmful to skin, and the more frequently you persevere in using it the worse the problems get.

Then look for alternatives (see below) to the array of aerosols, roll-ons and strongly scented cosmetics in your bathroom cabinet. Try in any case to stop buying the spray-cans (chapter four). And roll-ons tend to employ aluminium salts, which can be absorbed through the skin and add to your toxic load.

Alternative Toiletries

After-shave
Use diluted lemon juice, or mix 1 cup strong mint tea and 1 tablespoon vodka.

Talc Substitute
Arrowroot or tapioca or cornstarch.

Deodorant
Mix baking powder with a pinch of cornstarch or dunk a cotton wool ball in cider vinegar and rub where required.

Moisturizer
Pure almond oil, Simple, Roc, Clinique, coconut oil or olive oil.

Hair Conditioner
Beat one egg with half a cup of plain yoghurt. Rub mixture into washed hair, let stand three to five minutes, then rinse thoroughly with clean water.

Hair Lightener
Infuse two tablespoons chamomile flowers with warm water, two thirds of a pint.

Setting Lotion
Mix half a cup of warm water with three teaspoons of fine sugar.

Mouthwash
Gargle with mint tea or a teaspoon of baking powder dissolved in a glass of water.

Astringent
Use lemon juice diluted or grapefruit juice diluted.

Face Masks
3 heaped tablespoons cucumber pieces
1 egg white
1 tablespoon dried milk powder
Mix together, leave on 30 minutes, remove with lukewarm water, cleanse with water and moisturize

or
1 teaspoon honey
1 egg yolk
lemon juice, few drops
Spread on, leave a few minutes (maximum 15) and remove with lukewarm water, and then massage almond oil into skin.

Shaving Cream
Coat skin with soap or a light vegetable oil.

Toothpaste
1 part calcium carbonate (chalk)
1 part salt
few drops PURE peppermint oil (very expensive).
Mix together and store in a tightly fitting screw-top jar

or
baking powder

or
salt

All of these ideas work just as well for children too. They love to imitate and be included in their parents' routine. A hyperactive child will find cooperation difficult when he is irritable, but can take advantage of anything when he is well. Use bathing and breathing to reinforce his good spells.

Something else that children love is a sensitive soothing massage, and it very powerfully strengthens their sense of well-being and collectedness. Use a little almond oil and work over every part of his body with care and love. It is hard to do this badly, if your mood is right and you give yourself a little time. Do for your child what you would like for yourself. You will know by the result that you are on the right track – the child relaxes and breathes evenly, and you feel nicer too.

Banishing Lead

If your home was built before the Second World War or is on the fringe of an old neighbourhood, your water probably passes through lead pipes to get to you. It will pick up a surprisingly large proportion of lead while stagnant in the pipe overnight; even more if the water is soft and lathers easily. Provided the lead is near the house you can get rid of this by running off the cold water for two minutes every morning before using any of it. But in an old tenement or inner city area you should not trust this.

It would be better in the first place to use purified water BP for drinking purposes; it is available from the chemist. Bottled mineral water is an alternative, but usually more expensive. In the long run a filter jug such as the 'Brita' is usually cheaper. That will remove chlorine, copper, any particles and organic matter as well. Fluoride is not appreciably reduced. There are several more expensive systems available which vary in performance and price (see page 166).

The toxic effects of lead may be further offset by a diet rich in calcium and phosphate (see below), with moderate protein content and low in fat. Trace minerals such as magnesium, iron and zinc should be well catered for, and a multi-mineral supplement should contain in a daily dose at least 5 mg zinc for a child and 15 mg for an adult (see page 163). Wholemeal flour contains more zinc and nutrient minerals than white, and the fibre can also be somewhat protective against lead. Canned foods tend to contain lead; and all shellfish, liver and kidney are inclined to be contaminated during their growth. The above-ground parts of fruits and vegetables grown in areas of dense traffic are likely to be contaminated; but peas are protected by their pods and cabbage heart by the outer leaves, so these inner parts are safe. So are peeled fruit such as apples, especially because the pectin they contain tends to bind with lead and inactivate it.

CALCIUM AND PHOSPHORUS SUPPLEMENT
for Banishing Lead

Adult

calcium 700 mg daily, in 1,750 mg calcium carbonate.

phosphorus 500 mg daily, in 1,935 mg sodium dihydrogen phosphate.

bonemeal provides both, but is usually contaminated with
some lead from the bones.
Child
half the above.
Warnings
DO NOT continue beyond 1–2 months.
DO NOT take if you have had kidney or bladder stones

Glazed earthenware pots are lovely to display, but lead
leeches from the glaze into the contents if you cook or store food
in them. Aluminium pans are even worse and contain lead too.
So do hard PVC buckets and kettles. Glass and stainless steel
are fine.

If your suspicions about lead were well-founded, then the
people in your household affected by it should improve on these
measures quite quickly. You can then ease off the supplements,
but water filtration and avoidance of canned food would still be
wise.

Shopping for Clean Food
Unsprayed vegetables and fruit are a great asset. If your local
wholefood shop or greengrocer stocks them, buy there. But
there is little point in accepting unsprayed produce from a
display with traffic a few feet away. You simply trade lead,
formaldehyde and hydrocarbons for insecticides!

Among conventionally grown fresh goods, open lettuces and
cauliflower are badly contaminated right through. Unblem-
ished carrots are suspect; they tend to soak up insecticides in the
surrounding soil more especially than other roots. But the
skins and narrow root parts of parsnips, turnips and potatoes
can be just as heavily affected. Do not use these for stock or
soup.

Tight cabbage hearts and peas are well protected by their
natural outer casings, against sprays just as against lead.
Bananas and citrus fruit lose most of the contamination with
their peel. Apples and pears are likely to be penetrated for about
a millimetre beneath the skin.

Produce from third world countries, as stocked in shops
catering especially to Asian and West Indian tastes, is less likely
to be heavily sprayed than European produce. Many of these

items have thick unpalatable skins in any case, which protect the edible parts.

Try to find organically grown wholemeal flour, and rolled grains for breakfast cereals, to avoid the pre-harvest herbicides. And go for cellophane if you must buy cereals and vegetables pre-packed, because polythene and polystyrene leak gases into their contents. Cling film depends on its chemical coating to work, so be warned about that too.

Everything points back to good garden and allotment produce. If you can grow any yourself, make sure of the most suspect items first. Soft fruit, leafy vegetables and salads top the list, but composted unsprayed potatoes are a great boon if you can cope with them. Offer to buy an allotment-holder's surplus, if you trust his gardening methods. And make yourself known to any local wholesale organization dealing in organic produce. If you want to be sure of your winter supplies of onions, carrots and potatoes you need to be prepared to buy them by the sackful, in the autumn.

Tea and coffee are an active hindrance to cleansing, and it is worth knowing which herbs positively promote it. The cheapest is broom, but it tastes rather drab. Parsley piert has more flavour, but is much more expensive. Juniper berries make a tasty change, but need soaking overnight to get the best out of them. A teaspoonful of any of these (dried) makes a large cupful, after steeping for three minutes. Crush the soaked juniper berries before making them up with freshly boiling water.

Home Sweet Home

Now that you have worked through your food, water and toilet arrangements, give thought to the air in your home. Good ventilation cannot be taken for granted nowadays, especially in well-built and well-maintained homes. Double glazing and draught proofing can easily be good enough to spoil it, and stagnant air simply accumulates contaminants. See to it that some fresh air exchanges with your living space continuously. A brick vent is usually quite sufficient, or draughty window frames will do.

If a member of your family cannot tolerate tobacco smoke and the smoker cannot give it up, an ionizer can be installed in a room designated for smoking, and used whenever tobacco is alight and for some hours afterwards. This helps to destroy smoke which would otherwise hang in the room and eventually leak throughout the house. A candle will deal effectively with stale tobacco smells too, but may cause intolerance problems of its own.

The smoking-room should be ventilated to the outside, and be separated from the main parts of the house by a lobby, or 'air lock', if possible. All this is a good deal of trouble, but well worthwhile if smoking is important enough to be continued even when you know it harms another family member's health.

If you use air deodorizers, bleaches, spray polishes and carpet cleaners consider gas-free alternatives. See below for some simple recipes.

Alternatives for Cleaning Products
Air deodorizers
Fill large flat dishes or baskets with herbs, e.g. rosemary, thyme or marjoram or with spices, e.g. ground cardamom pods or coriander or cinnamon.
Use with pot-pourri and lavender.

Carpets
Get brand-new carpets steam cleaned with no chemicals to help remove dressings.
Spot clean with soda water or mix half-and-half white wine vinegar and water. Colour test first, out of sight.

Disinfectants
Soap kills germs, so substitute with soap. Boil to sterilize. Take better precautions – mark the crockery and cutlery of an infectious person with wool ties round handles, and keep separate. Flush toilets with hot soapy water. Flush drains with hot soapy water followed by a heaped tablespoon of baking powder or half a cup of vinegar.

Furniture polish
Damp dust and polish with pure nut oils, e.g. almond or walnut.
Remove 'rings' on furniture with a half-walnut rubbed into the mark.

Glass
Add a tablespoon of white vinegar to three cups of water. Use a flower spray and wipe dry.

or
Chamois leather well wrung out.

Clothes
Grate pure bar soap, add water then liquidize. Store in tightly fitting glass jar.

Surface cleaners
A mixture of half vinegar and half water will clean most things.

When clothes come back from dry cleaning, air them thoroughly out of doors, in the sun if possible, before hanging them indoors. That means choosing to clean them in good summer weather. Plan your interior decorating or timber treatment at this time of year as well. You need to ventilate the room very freely between coats of paint, and for some days after completion. The smell of new paint lasts for weeks otherwise.

Sickening Buildings
When planning the kind of home improvements that will last a lifetime, think very carefully before choosing the cheap option. You generally get the quality you are prepared to pay for.

Avoid plastic glazing if you can. Gases vapourize from it, and large areas of modern windows heating in the sun can yield appreciable amounts of vapour into the room. Glass is much more inert. Urea-formaldehyde foam cavity wall insulation gives off gas badly in some circumstances. Choose solid flocking, if you plan this operation.

Make sure that heating apparatus is well ventilated to the outside and that all flames burn cleanly. Install an exhaust hood over a gas stove if at all practical, or place an extractor fan nearby. If you plan to install a central heating boiler, site it outside the house if possible. And make sure flues vent well away from windows which open – more if possible than the minimum required.

If you have a child very intolerant of chemicals, you need even to consider the building materials of which your home is constructed. Laminated wood, chipboard, plaster and resin glues gas off formaldehyde for a long time when they are newly installed. Vinyl floor coverings smell new for some time too, and resins from softwood are quite troublesome. The safest materials are stone, ceramics, hardwood and steel. Even aluminium is more troublesome than hardwood.

Inside the Home

Most people nowadays buy artificial fabric garments for economy and wear, and if you are among these you do not have to change your wardrobe completely overnight! But if you are chemically sensitive you will do better to choose cotton, wool and linen as you replace worn items of clothing. Remember that linings need checking as well as the main fabric of the garment.

Beware in any case of special fabric treatments, whether for permanent pressing, crease resistance, water repellence, flame-resistance, moth-proofing or shrink-proofing. All these processes involve formaldehyde, which is a highly sensitizing vapour. Fabrics are by no means the only way we are exposed to it, but are the most intimate.

Curtains, cushion covers and upholstery need to be considered in the same way. Newly tanned leather goods will vapourize formaldehyde for some time – think twice about that Chesterfield! Beware of cheap expanded polyester stuffings for it too; they can release irritant vapours for a large part of their lives. Feathers and horse-hair are preferable; if you are allergic to these, you can probably be desensitized. Otherwise you will need to cover these natural, gas-free stuffings with several layers of a fabric you get on with.

When the time comes to change carpets, try to afford wool and avoid foam backing. This will matter less to you if no one is really sensitive to chemicals. But the large area of a carpet, and the fineness of the fibrils in an artificial pile, mean it can go on yielding solvents and insecticides in sensitizing quantities for a large part of its life.

Artificial pile carpet will also deaden the electrical quality of the air in the room very drastically (chapter four). Good ventilation, natural fibre furnishings and conventional lighting minimize this effect; fluorescent tube lighting, television and video screens, air conditioning, radiant and fan heaters make it worse. If they are unavoidable, or smoke fumes and dust are a nuisance, an ionizer is a very worthwhile investment. It will filter the air and improve its electrical quality in one operation.

The dryness of central heating can be a problem. It is improved by fresh flowers and potted plants in the room – provided they are kept watered! Humidifiers on the radiators may even be necessary; check first how rapidly a shallow dish of water evaporates dry. Excessive condensation in old buildings can, of course, bring decorative problems, but only impair health if moulds grow. Fungal spores are very fine and waft about the room under certain conditions, and are very irritant if breathed. Anyone allergic to them will have serious trouble in a mouldy bedroom. Check behind large items of furniture if the room smells musty; moulds can very often grow to large colonies before you realize they are there.

Lastly, bear in mind electronic apparatus and the magnetic fields associated with electric motors and transformers. If you have someone sensitive to these, you will probably succeed in confining their worst effects to rooms he need not use. And if you have carefully attended to everything else you reasonably can, his tolerance will usually be reasonable too.

The Garden

Out of doors, you will think twice before spraying your garden with anything, avoiding pesticides and using 'softer' alternatives. See Further Reading (p. 193) for guidance on what to choose.

Creosote the fence, or use any other kind of wood-stain or preservative on airy days. Most of these have adequate warnings on their labels: take them seriously. Accommodate animals elsewhere too, while the fluid dries.

Unless a garage is exceptionally well ventilated, the fumes that accumulate while a car is parked or started up are an appreciable hazard to anyone, sensitive or not. If you have the choice, a car port is preferable; ideally it should be sited away from the house.

Nobody seems to cut their lawn by muscle-power any more, and anyone using a motor mower would be wise to choose a breezy day. The same applies to heavy power tools.

With average good fortune most of these considerations do not apply urgently to you. Only a very few will be faced with working laboriously down the list, and those will probably seek expert help in any case.

The principal lesson for most of us is how many beguiling choices we can make in innocence, landing ourselves with consequences we may have to put up with for the rest of our lives. A great many new fabrics and materials are formulated every year, and introduced too rapidly for us to know with any certainty how they will impinge upon us. Naturally, where they seem to offer a convenience or economy we have previously been denied, pressures from producers and consumers tend to coincide. It is unreasonable of us to blame the manufacturer, if we rushed willingly to patronize his less expensive product in the first place. No consumer law is absolutely proof against the folly of consumers.

So it is up to us all in the first place to be cautious. Then we must be prepared to put our money, and our influence, where our suspicions are.

CHAPTER FIFTEEN
Taking Part

It will take a lot more than one reading of this book to come to terms with what we have to say. It depends on how much of the truth you knew already. But you are unlikely to have read this far unless your experience bears us out in some degree. If so, we hope we have been helpful to you in some practical way.

But with the realization of what is happening in the world, dawns responsibility. Everyone who understands is in a position to help change things, and no one else is. The first danger is that you will give way to a sense of urgency. Nothing would defeat your efforts sooner. The power of your enthusiasm will exhaust you, and engender resistance in the very people who stand most in need of what you can offer them. Bide your time, conserve your energy; let the situation come home fully to you first, before you do anything. There is a lot else to read and to discover while you are waiting to see what will happen next. We have listed in Further Reading a selection of books which will widen the scope of your insight. Not every one will appeal to you, but several of them will.

And you have seen scope for consolidating the goodness of life in your own home. That is something to savour well. It may be some years since you felt any real sense of purpose in the day-by-day tasks of housekeeping, gardening, shopping and cooking. Yet they are an essential foundation on which to base any quality life is to have. The effort they take need not be large, but must be carefully and knowingly applied. It is in doing well the simple things that feeling well arises.

It may come as a shock to realize how you may actually be undermining your general health through your occupation. That prompted me to change my own work, and set me a difficult and exciting personal journey which continues.

But I have more scope for choice than most, a privilege I am constantly aware of.

Livelihood is necessary, even if it cannot immediately be improved. Yet there is none which gives no opportunity of improving life. Dwell not on the work you are obliged to do, but on your options in the way you do it. Just as at home, quality hinges on the manner and care with which you do what must be done. No one makes peace or maintains good cheer by accident, and dozens of opportunities to do so arise in every working day. No task is so repetitive or meaningless that you cannot benefit by your attention to it. To be able to devote yourself to whatever task you have in hand is a faculty worth learning.

Then in the long run, chances will arise to change your work. Bureaucratic and mechanistic values are breaking down, and will eventually give place to more personal and quality-conscious forms and scales of organization. The minority already conscious of these needs and motivated to meet them are tuned to spot the opportunities. They will become the people who create new forms of work which will replace the old. That takes the kind of personal stature anyone can grow into, given the will and the time. My guess is you have both.

All that will have saved you from becoming a bore, and got you listening instead. It is fascinating to consider where other people really are in their lives. The things you notice first are the ones you just stopped doing – driving miles for a penny off a gallon of petrol, or queueing for hours to get into a sale; complaining about the weather, or the government; and chatting for ages on telephones and street corners, saying nothing worth listening to and hearing even less.

It took me years of listening to discover which questions people want the answers to, and how to feed them ideas and information they are ready for. I still make mistakes and get impatient. Yet I know perfectly well that people cannot be hurried, and will not be bullied. And surely as I give up pressing on them what they need, they start asking for it.

To help people discover what you know already is necessary and worthwhile. But you need not do anything for that to happen. You have only to be the right person, and to know. Let neighbours, friends and relatives approach you, in the round-about way that people often will. They may ask you questions, and seem only casually interested in your answers. Speak carefully of what you know from personal experience, and do not generalize or make bold claims. Your questioner admires you shyly, and means to follow your example in some matter of her own. Answers from experience cannot mislead her. If she is bolder and asks your advice openly, answer along the lines, 'This may not suit you of course, but what I did with my children was . . .' That will keep you on safe ground. Never be afraid to share what you know with people who are interested. But always beware of acting as their guide. Encourage them by all means to explore for themselves answers to their own problems. Only avoid offering them expectations, and you cannot disappoint them.

Telling

The first person you may wish to influence is your doctor, particularly if he laboured mightily and in vain to achieve with drugs what you yourself have managed with common sense.

I recall a man who undertook a naturopathic treatment for expelling the gallstones his conventional doctor had discovered. It was not comfortable medicine, but at the end of it he passed twenty nuggets which looked as if they might be gallstones. We had one analysed to prove it. Triumphant, he went and saw his doctor waving the bottle of stones, and asked if he might have another X-ray just to prove that all the stones had gone. His doctor hit the roof and showed an astonished patient the door!

I do not happen to think there was any way that man could have had his wish except by paying for it. He was challenging the basis of the doctor's self-respect, puncturing it like a balloon. Even the greatest of tact could not disguise that, the way it happened. But he could have set about the whole plan differently. Suppose he had taken the doctor into his confidence in advance, showing every respect for his advice and opinion but indicating firmly his own wish and willingness to try this non-operative treatment? He might then have managed to engage the doctor's curiosity, and help him to feel that he was still in charge. At least he tried. Anything is better than letting the false notions of a profession survive unchallenged. Its members may not like being presented with evidence of that falsity, but it sinks in eventually. The knowledge of it is slowly digested into perspective, and may show many months or years afterwards as an alteration in attitude or judgement.

If you want to do better than my friend with gallstones, pay your doctor a courtesy call to discuss with him an exclusion diet, or any other plan you have in mind. Secretly you may already have put it into effect, but it is better not to admit this. Ask his advice and permission to undertake it, and promise to let him know the results. Then remember to do so; not at great length, and without exaggeration. A letter might be best, presented in person during a short consultation for him to read afterwards. Enclose details of where you got the diet from. And remember to thank him sincerely for his advice.

Another good approach is to reconsider the kind of questions you are asking the doctor about a complaint that keeps on troubling you. He may feel under pressure to do something effective every time you see him, and so to keep on offering a treatment that you know already does not work.

Change your approach completely. See him by appointment sometime when you are not actually suffering from the complaint, and make that clear from the start. Say that you appreciate his advice and help, but feel you are wasting his time by letting the complaint keep recurring. Is there not something you can do to help prevent each relapse? Are there any diet supplements that might strengthen your resistance? Is there anything you should avoid? Are any brands of medicines available without colourings? Say that you realize he may not have answers to these questions off the cuff, but that you will willingly return after an interval to receive his considered advice.

Then, if nothing comes of your offer, you can feel free to write to him later about anything you have discovered, and where he can read about it. This is bound to interest him if by then you have stopped attending his surgery repeatedly for the same thing. There will be several others with a similar problem who still do!

A third approach requires a little more skill in dealing informally with people, but can work very well if you have this gift. Talk to the doctor's wife, in the shops or at a social gathering. Sometimes a receptionist or nurse, or the health visitor, will be a more practical route to his ear. Make it clear that the message is friendly, and do not be too direct or obvious about its content.

A health visitor is often a good listener in her own right anyway, and may or may not be on close terms with the doctor. She can make use of your experience directly in her own work, however, and may even ask you to help someone else try it. As a professional group, health visitors seem to have fewer inhibitions about new ideas and are more likely to encourage your initiatives.

You need not be so subtle with shopkeepers. Tell them straight out what you would like, and if they do not have it do not buy. Write to the marketing director of any supermarket chain you care to, in completely positive terms. Say what you would like to be able to buy and why. Comment on their particular lines or labels if you have observations to make, briefly and to the point. You can be sure your influence will be felt.

The directors of supermarket chains and food processing and distribution organizations are intelligent people. They do not necessarily want to over-rule your preferences; catering for firm trends is in their own best long-term interest. They are ready to move their resources wherever the market is most promising. When they do something you can support, make good use of it and let them know you approve. Be careful to budget your letters, however. It does not help to be too clearly identifiable as a regular correspondent. Maybe a friend will write next time instead.

Letters to local district councils and Members of Parliament carry little weight on consumer matters, where the main decisions are commercial. But fluoridation of the local water supply is an issue for your local health authority. And the water authority should take an interest in any instances of pollution to drains and watercourses that you may spot. Incidents in which people or gardens are sprayed accidentally by aircraft should be reported in the first instance to the local police, but do try first to obtain photographs if you have the chance. Verbal estimates of distance and height by the victims of such incidents tend not to be credited by investigating officers; any hard evidence you can produce will strengthen your claim.

Campaigning

The best way to influence Members of Parliament and government bodies is as a member of the special interest groups who most closely represent your interests. Join them, to keep in touch with developments and opportunities nationally. They will be able to alert you to issues as they come up, so that you can write to your representatives giving your views just when they are of most use. Use your own words, and write as a private citizen. Never address letters to ministers direct, but to your Member of Parliament. When an MP forwards your letter it must come to the attention of a minister; letters direct are normally processed by civil servants.

These are the special interest groups we know about, which cover most of the issues raised by the subject of this book. They are in alphabetical order.

Action Against Allergy
43 The Downs,
Wimbledon,
London, SW20.
Tel. 01-947 5082

A self-help and campaigning group who represent the interests of sufferers from allergy.

Environment for Hypersensitives
c/o 5 Kilham,
Orton Goldhay,
Peterborough,
Cambs. PE2 0SU.

A group attempting to establish a holiday home catering for severely hypersensitive people.

Environmental Medicine Foundation
111 Toms Lane,
Kings Langley,
Herts.

A charity constituted by a number of clinical ecologists to promote research into allergy and its treatment.

Friends of the Earth Limited
377 City Road,
London, EC1V 1NA.
Tel. 01-837 0731

A vigorous and well-informed organization campaigning effectively on all environmental issues, with many local groups.

Foresight
The Old Vicarage,
Church Lane,
Witley,
Godalming,
Surrey, GU8 5PN.
Tel. 042879 4500

A group concerned especially with health in the pre-birth period, who have by extension become involved in self-help for hyperactive children.

Henry Doubleday Research Association
National Centre for Organic
   Gardening,
Ryton-on-Dunsmore,
Coventry, CV8 3LG

Of special assistance to novice organic gardeners, offering a wide range of very practical literature and goods.

Hyperactive Children's Support Group
59 Meadowside,
Angmering,
Littlehampton,
Sussex, BN16 4BW.
Tel. 090 62 70360
Office hours: 0903 725182

A very successful and well-established mutual self-help group for parents of hyperactive children. They offer practical advice and literature and sponsor research, as well as campaigning effectively on relevant issues alongside other organizations.

National Anti-Fluoridation Campaign
36 Station Road,
Thames Ditton,
Surrey, KT7 0NS.
Tel. 01-398 2117

A small but energetic group who make known to public health authorities and MPs the case against fluoridation of water supplies. A useful range of publications, though in a rather polemical style.

The Soil Association Ltd.
86 Colston Street,
Bristol, BS1 5BB.
Tel. 0272 290 661

The consumer organization for organic agriculture, market gardening and horticulture. They publish a quarterly magazine, and a range of literature including an inexpensive classified list of food additives.

Templegarth Trust
82 Tinkle Street,
Grimoldby,
Louth,
Lincs. LN11 8TF.
Tel. 0507 82 655

A registered charity
researching the nature of
health, and how it can be
cultivated directly in modern
circumstances. They are
actively promoting
health-oriented change in the
East Midlands, and publish a
newsletter and a small range
of literature. Endorsed by the
World Health Organization.

All of them will welcome your interest and support. We have
not omitted any group on purpose, and have no doubt there are
many worthy groups we do not yet know about.

Getting on with Gaia

How much you hope for from efforts like these will depend on
where you place yourself in some ultimate scheme of things.
Considerations like that have not been fashionable for some
time. Organized churches and religious groups have not in
general kept their thinking and behaviour abreast of man's
needs in this century. Rather unexpectedly, new breeds of
physicist, ecologist and biologist are beginning to occupy the
vacancy.

Their work suggests that life on earth functions as one great
organism, which has maintained itself throughout many changing fortunes for at least a quarter of a billion years. We met her
in chapter seven, and called her Gaia. Does what we have seen
happening to our children make sense as Gaia's behaviour?

Questions of proof do not arise. We are concerned only with
the usefulness of one concept or another; with how well ideas fit
our experience, and where they may lead us. If the world is a
chemical machine, then it is silting itself up and breaking down.
There is no reason to expect prosperity and happiness in that
paradigm, even for privileged minorities. There is no sign that
the people most devoted to this idea are healthier for it, and
some indications of the opposite. But if instead we are all cells in
the body of one global being, the facts make better sense.

Materialist behaviour makes man's settlements stand out

against the rest of nature. Under a large enough microscope they would resemble cancer in the tissues of a human, malignantly destructive and a serious threat to life. Gaia's defences react as ours would do, attacking the malignant tissue through its metabolism, inflaming most acutely against the greatest threats. These come from ourselves – the industrialized communities who have most thoroughly disrupted nature's pattern. Gaia's inflammatory defences reach us through the ordinary habits of Western life.

What in human disease takes months, in Gaia is taking a hundred years. Earlier in this century we began to experience new epidemics of degenerative disease in generations first exposed late in life to the consequences of food processing. Those and more recent epidemics are now penetrating to younger people, the most vulnerable human cells in Gaia. Lives which should flow easily are frustrated by stubborn resistance they cannot understand. It seems unfair that they should suffer for a trend they did not start, generations after it began. Yet nothing less could ever have undermined their parents' confidence sufficiently to make them question its validity. Many have consequently become active leaders in a campaign for wholesome change.

Chemically sensitive children and their families are pioneering a trail back to health for the rest of us to follow. When we understand that, most of us willingly accept their lead. Few rate an obsolescent lifestyle so highly that they will sicken rather than lose it. Meanwhile those who have regained their senses are beginning to be taken seriously. Naturalistic movements in agriculture, medicine and education are gaining ground, as increasing numbers of us rediscover the allure of health. The human fabric of Gaia is beginning to heal. That is the real work for people willing to take part – securely knitting up the scars.

Further Reading

Chapter One

Rachel Carson, *Silent Spring*, Penguin, 1965.
Ben Feingold, *Why Your Child is Hyperactive*, Random House.
R. Mackarness, *Not All In The Mind*, Pan, 1972.
R. Mackarness, *Chemical Victims*, Pan, 1980.
T. Randolph, *Allergies, Your Hidden Enemy*, Turnstone.
Hans Selye, *The Stress of Life*, McGraw Hill, 1956.

Chapter Two

Lady Eve Balfour, *The Living Soil*, Faber, 1943.
R. Van Den Bosch, *The Pesticide Conspiracy*, Prism, 1980.
Dr D. Hodges, *Agriculture, Nitrates & Health*, Soil Association
    Quarterly Review, September 1985.
H. H. Koepf, *Nitrate, An Ailing Organism Calls for Healing*,
    Biodynamics, No 73, 1965.

Chapter Three

Sir A. Howard, *An Agricultural Testament*, OUP, 1940.
Ivan Illich, *H_2O And The Waters of Forgetfulness*, Resurgence,
    1985.
Prof F. King, *Farmers of Forty Centuries*, Rodale Press, 1911.
R. McCarrison, *Nutrition and Health*, McCarrison Soc, 1982.

Chapter Four

The Gaia Atlas of Planetary Management, Pan, 1985.

Chapter Five

Walker & Cannon, *The Food Scandal*, Century, 1985.

ed, F. Lawrence, *Additives, Your Complete Survival Kit*, Century, 1986.

Chapter Six

M. & S. Crawford, *What We Eat Today*, Neville Spearman, 1972.
F. Pottenger, *Pottenger's Cats*, Price-Pottenger, 1983.
O. Schell, *Modern Meat*, Random House, 1984.
H. Schroeder, *The Trace Elements and Man*, Devin-Adair, 1973.
H. Tomlinson, *Aluminium Utensils & Disease*, L. N. Fowler, 1958.

Chapter Seven

J. Lovelock, *Gaia: A New Look At Life On Earth*, OUP, 1979.
K. Pedler, *The Quest for Gaia*, Granada, 1981.
P. Russell, *The Awakening Earth*, Ark, 1984.

Chapter Twelve

S. Lewis, *Allergy? Think About Food*, Wisebuy, 1986.

Chapter Thirteen

Barnes & Colquhoun, *The Hyperactive Child*, Thorsons, 1984.
Grant & Joyce, *Food Combining for Health*, Thorsons, 1984.
L. Hills, *A Month by Month Guide to Organic Gardening*, Thorsons, 1983.
L. & S. Kenton, *Raw Energy*, Century, 1984.
L. Kenton, *Ageless Ageing*, Century, 1985.
Dr C. Pfeiffer, *Zinc & Other Micronutrients*, Pivot Original Health, 1978.

Chapter Fourteen

T. Heinl, *The Baby Massage Book*, Coventure, 1982.
F. Leboyer, *Loving Hands*, Collins, 1977.
L. Kenton, *Ageless Ageing*, Century, 1985.

Index